Women in Engineering and Science

Series Editor
Jill S. Tietjen
Greenwood Village, Colorado, USA

More information about this series at http://www.springer.com/series/15424

Debra A. Christofferson
Editor

Women in Security

Changing the Face of Technology and Innovation

Springer

Editor
Debra A. Christofferson
Tempe, AZ, USA

ISSN 2509-6427 ISSN 2509-6435 (electronic)
Women in Engineering and Science
ISBN 978-3-319-57794-4 ISBN 978-3-319-57795-1 (eBook)
https://doi.org/10.1007/978-3-319-57795-1

Library of Congress Control Number: 2017942744

Printed on acid-free paper

This Springer imprint is published by Springer Nature
The registered company is Springer International Publishing AG
The registered company address is: Gewerbestrasse 11, 6330 Cham, Switzerland

Foreword

Our personal and professional worlds are increasingly digital, and the information we share in this digital economy is proliferated across a multitude of networks. These networks have become so ubiquitous that we often engage with them without deliberate thought as to the information we provide. Whether playing online games, building our social network, conducting financial transactions, or shopping online, each time we engage with a network we provide another block of information. Each block of information on its own may seem innocuous; however, hackers have built a prosperous business by obtaining and consolidating disparate blocks of data into sensitive, saleable information.

While the risks with our personal information are immense, the risks to businesses are astronomical. In addition to maintaining legal compliance, ensuring the security of the business network is central to safeguarding a company's intellectual property, reputation, and relationships with its customers and suppliers. This is particularly germane in supply chain as suppliers are often the gateway into large organization networks. We've all heard countless examples of security breaches that originated in the supply chain, including the breach of Target's network which resulted in the theft of 40 million credit card numbers. Then there was the theft of 21.5 million records with highly sensitive personal information from the U.S. Government's Office of Personnel Management, and the pre-installation of a malicious app to steal passwords and credit card information on Android devices. In each of these cases, the hackers gained access through a supplier's network. Cyber breaches, such as these, have caused significant financial harm to businesses, from disrupting operations to extortion to industrial espionage. The reputational damage can be so severe that many businesses do not report breaches.

Without a doubt, such breaches have a drastic effect on consumers and businesses—with both short and long term effects. As society's dependence upon digitization increases so does the importance of cybersecurity. We're moving into a space of autonomous vehicles, increased internetworking of devices through adoption of the Internet of Things (IoT), wearables, and blockchain for records management and payment—all of which rely on sharing information across copious networks. The migration to cloud services, integration of suppliers, and implementation of applications to analyze big data have resulted in the centralization of data. This is a huge target for hackers as it provides unparalleled access to massive amounts of data with just one attack. Thus, the protection of our networks is crucial.

We are reaching a tipping point. With the increased interconnectivity and consolidation of vast amounts of data, we're likely to experience an increase in cyber breaches. At the same time, we have a global shortage of professionals entering the cybersecurity field. The 2015 (ISC)2 Global Information Security Workforce Study[1] projected a shortfall of 1.5 million workers in cybersecurity by 2019. The study also showed that women comprised a **mere 10%** of the cybersecurity workforce globally.[2] Diversity is key to innovation, thus attracting women to this field is vital. In addition to diverse thought, women bring an inherent ability to collaborate, multitask, empathize, and take a holistic approach to problem solving. Since data is spread across multiple nodes and affects a wide range of people, protecting networks often requires bringing many organizations together to employ a holistic solution and secure our data. We need the skills that women bring if we are going to successfully address cybersecurity risks.

The scarcity of role models and mentors is one of the key obstacles in attracting women to this field. Several organizations have worked tirelessly to increase awareness of this field and attract and advance women in cybersecurity. This book is a milestone. The women are accomplished, talented, fierce, and, above all, excellent role models. The background of these innovative women in cybersecurity and their depth of knowledge will inspire other women. I encourage you to read this book, be inspired, and take action to address this important topic.

M.L. Peck

[1] The 2015 (ISC)2 Global Information Security Workforce Study, Frost & Sullivan, April 16, 2015, https://www.isc2cares.org/uploadedFiles/wwwisc2caresorg/Content/GISWS/FrostSullivan-(ISC)²-Global-Information-Security-Workforce-Study-2015.pdf

[2] *Women in Security: Wisely Positioned for the Future of InfoSec*, Frost & Sullivan, 2015, https://www.isc2cares.org/uploadedFiles/wwwisc2caresorg/Content/GISWS/2015-Women-In-Security-Study.pdf

Contents

About the Editors

Juanita Agard was born in New York City, NY, and raised in the Seattle, Washington, area. She moved to the nation's capital to attend Howard University and pursued a bachelor's degree in System and Computer Science with a minor in Mathematics. Immediately after graduating, Juanita had an IT Specialist position with IBM waiting for her, and the experience confirmed her passion for entrepreneurship, Information Technology (IT), and technology law.

As cybersecurity emerged, it eventually developed its own career path, along with educational institutes creating cybersecurity curriculums. Juanita envisioned uniting her IT engineering experience with technology law to become a cybersecurity subject matter expert. She researched cyber education partnerships between companies and universities and job opportunities with companies offering cyber training.

Juanita landed a Software Requirement Engineer position with Booz Allen Hamilton, a company with a reputation for investing in their employees. Through the Booz Allen Cyber cohort program, Juanita immediately enrolled into the Cybersecurity Technology master program at UMUC and expects to complete it in 2017.

Ms. Agard has been with Booz Allen Hamilton for almost 4 years, majoring in a cybersecurity master's degree program at UMUC. She also obtained an **active security clearance** and is gaining all the cybersecurity experience she can.

In parallel, Agard too took the CompTIA Security +, and EC-Council Certified Ethical Hacker (CEH) boot camp trainings to obtain her cyber certifications. She is **studying for the CISSP** and planning to complete it this year. Ms. Agard also obtained a Cybersecurity Technology Graduate Certificate and has only two classes left to complete her master's in cybersecurity.

She currently is working at NASA as a cybersecurity policy analyst. Juanita is relatively new to the project and still on the learning curve as to how things operate. Her responsibilities are to capture the existing security change management process from multiple projects, to develop a high-level Change Control Board charter. The overall goal is to have an overarching mechanism to manage IT security services agencywide.

On a previous project, she served as a Security Requirement Analyst, analyzing and consolidating NIST SP 800–53 security controls to import into a central repository for reusable and repeatable purposes for the FDA project.

Diane Barrett holds a PhD in business administration with an information security specialization from Northcentral University. She is a Certified Information Systems Security Professional (CISSP) holding many additional industry certifications. Diane has an extensive background and has been involved in the IT industry for over 20 years. She is the President of NextGard technology, LLC and has done contract forensic and security assessment work for numerous years and held positions such as manager of research and training for Kroll's cyber division and forensic training director for Paraben Corporation.

Diane is the conference program chair for the Conference on Digital Forensics, Security and Law as well as the President of the Digital Forensics Certification Board. She has been involved in collegiate-level forensic education at Bloomsburg University, American Military University, and the University of Advancing Technology. She has co-authored several security and computer forensics books including *Security + Exam Cram, Virtualization and Forensics, and Cybercrime and Cloud Forensics: Applications for Investigation Processes*.

Lacey Chong is a national security consultant who specializes in business process improvement. She has over 13 years of experience working for the U.S. Department of Defense, U.S. Department of State, and several national-level intelligence agencies as a government civilian and as a consultant. Throughout her career, she has held various roles involving cybersecurity, project management, all-source intelligence analysis, and business process improvement. Most notably, she served as a Special Assistant at the National Security Council's Counterterrorism Policy Directorate from 2008 to 2011. Ms. Chong is currently an Associate with Booz Allen Hamilton in McLean, VA.

Ms. Chong holds a B.A. in Asian Studies from the University of Puget Sound, an M.A. in International Affairs/International Security from George Washington University, and a graduate certificate in Change Management from the Georgetown University Executive M.B.A. program. She founded Spa Swag for Warriors, a non-profit charitable organization in 2015, and is currently a member of the Board's

Executive Committee. Selected as a 2016 Northern Virginian of the Year for her work with Spa Swag, she directs strategic communications, operations, and outreach efforts on behalf of the non-profit. She is an avid traveler, yogi, and foodie. Ms. Chong lives in Sterling, VA, with her husband, adopted dog and cat.

Debra Christofferson has more than 25 years of IT and security management experience, in a Fortune 500 environment, across the United States, Europe, and Asia, with Intel Corporation, the Apollo Group, and the State of Arizona Security and Privacy Office. She has supported security often from the ground up, and worked across multiple lines of business in the public and private sectors in consulting roles. Ms. Christofferson currently serves on the Information Systems Security Association's International Board of Directors and chairs the CISO (Corporate Information Security Officer) Advisory Council for ISSA's CISO Executive Forum. She also leads the local chapter for the Cloud Security Alliance that she co-founded. Christofferson holds the CISSP (Certified Information Systems Security Professional) and CISM (Certified Information Security Manager) security certifications and facilitates local chapter CISM workshops.

Martha Daniel is the founder, president, and CEO of Cytellix and its parent company, Information Management Resources, Inc. (IMRI), responsible for successfully deploying network security and asset management solutions to local, national, and international organizations of every size in a wide range of industries including financial services, healthcare, biotech, education, logistics, and manufacturing. A fearless entrepreneur with 35 years of technology expertise and unsurpassed knowledge of program management and enterprise IT solutions, she is driven by a vision to translate business needs into leading edge technology solutions that help protect customers in both the private and public sectors throughout the world.

In addition to launching Cytellix, Daniel has led IMRI in becoming an industry frontrunner in cybersecurity, program management, and engineering services for federal and commercial entities, employing more than 155 professionals worldwide and managing over $300 million in data center operations located in 19 U.S. states. Clients include the U.S. Department of the Navy, U.S. Army, and U.S. Air Force, as well as small and mid-size businesses and Fortune 500 companies such as Wells Fargo, Lockheed Martin, and IBM. Prior to IMRI, Daniel had a successful corporate career, serving as chief information officer at FDIC/Resolution Trust Corporation and senior systems engineer at IBM. She also proudly served as a cryptologist in the U.S. Navy. As *the* unsurpassed authority on the subject of cybersecurity, Daniel's views are regularly profiled across digital, print, and television mediums. She is coauthor of *The Other Side of Midnight, 2000,* has written for *The White House Blog,* and was recently selected as a study group member for the National Infrastructure Advisory Council's (NIAC) Water Sector Resilience Final Report and Recommendations.

Daniel's diligence in inspiring and supporting success in the community has led to keynote speaking engagements at the 6th annual Women as Veteran Entrepreneurs, Irvine Valley College 9/11 Commemoration, and the Biola University's Crowell School of Business MBA Distinguished Speaker Series.

The U.S. Small Business Administration, Santa Ana District Office, selected Daniel as the 2016 Small Business Person of the Year. She was also a recipient of the 2015 Patriot Award given by the Employer Support of the Guard and Reserve (ESGR), established by the Department of Defense. Alongside Chick-fil-A CEO, Dan Cathy, and others, Daniel was honored by the Passkeys Foundation as a 2014 Leader of Integrity for American Life Technology. Additionally, she was honored as one of the top women veteran leaders for the 2014 White House Champion of Change.

A true advocate for change, Daniel contributes her time and resources to various community organizations including the Rothenbuehler Foundation for Veterans, New Directions for Women, the Child Guidance Center, and the Vicksburg Soccer Organization. In addition to being a contributor and guest speaker at the 39th Annual Economic Forecast at Chapman University hosted by James L. Doti, Ph.D., Daniel is also an ordained minister with the African Methodist Episcopal 5th District, a member of the Trusteeship-International Women's Forum, a Vice Chair of the Orange County Business Council Cybersecurity Task Force, and a member of the Orange County Homeland Security Advisory Council. She holds certifications in IBM Project Management and Contract Quality Assurance. A graduate of California State Polytechnic University with a degree in computer information systems, Daniel earned her MBA from the University of La Verne, California.

Mary Ann Davidson is the chief security officer at Oracle, responsible for Oracle software security assurance. She represents Oracle on the board of directors of the Information Technology-Information Sharing and Analysis Center (IT-ISAC) and serves on the international board of the Information Systems Security Association (ISSA). She has been named one of *Information Security*'s top five "Women of Vision," is a Federal 100 Award recipient from *Federal Computer Week*, and was recently named to the ISSA Hall of Fame.

Davidson has served on the Defense Science Board and was a member of the Center for Strategic and International Studies Commission on Cybersecurity for the 44th Presidency. She has testified on cybersecurity to the US House of

Representatives (Energy and Commerce Committee, Armed Services Committee, and Homeland Security Subcommittee on Emerging Threats, Cybersecurity, and Science and Technology) and the US Senate Committee on Commerce, Science, and Technology.

Davidson has a B.S. in mechanical engineering from the University of Virginia and an M.B.A. from the Wharton School of the University of Pennsylvania. She received the Navy Achievement Medal when she served as a commissioned officer in the US Navy Civil Engineer Corps.

Rhonda Farrell is an Associate with Booz Allen Hamilton, primarily focusing on cybersecurity related life-cycle activities within the IC, DoD, and Federal civilian markets. She has led design, quality, and cyber-assurance activities for the federal intelligence and DOD communities aligned with research and development labs, cybersecurity operations centers, and technological infrastructure program management offices. Corresponding areas of focus have been on the introduction of change and quality management principles, process improvement, and supporting strategic planning, risk management, and communication programs. Her prior career experience included operations, engineering, and security functional areas of Fortune 500 companies throughout Silicon Valley, CA, as well as with the U.S. Marine Corps at Quantico, VA.

Her educational background includes a B.S. in Business Administration (heavy Math and C.S.) (1999) and an MBA in Strategic Management (2000) from California State University; a J.D. with a Technology focus from Concord Law School (2009); and is currently preparing for her dissertation defense for her Doctorate of Science in Information Assurance from the University of Fairfax.

She is a veteran committee member and leader within IEEE Region 2, as well as the Northern Virginia and Silicon Valley, CA sections where she has developed initiatives focused on increasing member value, professional development opportunities, section growth, and realization of strategic partnering opportunities. She brings to the section leadership team enthusiasm, a strong work ethic, commitment to IEEE and team-oriented principles, a deep technical background, sound management capabilities, and a solid grounding in cyber and quality best practices.

Rhonda also serves on the International Board of Directors for ISSA—the Information Systems Security Association. She is founder of the Women in Security Special Interest Group (WIS SIG) and leads the SIGS on behalf of the Board. Rhonda is an ISSA Fellow.

Miriam Fernandez currently supports security platform development with Intel Corporation as an Information Security professional in Intel's Key Generation Facility. With 25+ years of experience, Miriam executed multiple IT roles with a global telecommunications and service provider-- Hewlett-Packard (formerly Alcatel Lucent, Lucent Technologies, and AG Communications) for the Americas region. She is an experienced information systems engineer supporting large enterprise IT data centers and networks. These include UNIX servers, and leadership of disaster recovery and business contingency planning programs. She holds a bachelor of science in Computer Information Systems, and earned a Six Sigma Green Belt. Miriam currently serves the Cloud Security Alliance for the local chapter board in AZ and previously was on the leadership teams for the Information System Security Association's (ISSA) Phoenix chapter, and the Sonoran Desert Security User Group (AZ SDSUG).

Pamela Fusco has accumulated more than 30 years of experience as an industry executive and expert in cybersecurity. Ms. Fusco has organized and managed multibillion dollar global strategic and tactical business driven architecture, technology, and compliance initiatives. She has led numerous global teams to implement innovative business enabling platforms.

In the past, Fusco resided on the U.S. Presidential White House Inaugural Staff. She has held positions as the chief security strategist; chief information security officer roles with the Apollo Group, Merck & Co., Inc., Digex Inc. (MCI Security Solutions); and EVP of global information security at Citigroup. Pamela was an initial founder of SAFE Bio Pharma Inc., serving on its board of directors.

Fusco began her career in the U.S. Navy serving as a cryptologist, where she supported security proceedings for government and national intelligence and SPECOPS (special operations).

Ms. Fusco has been bestowed with numerous awards and honors including a Presidential Citation, Information Systems Security Association (ISSA) International Hall of Fame, Distinguished Fellow, and Woman of Influence. Fusco is CFO (Chief Financial Officer) of the ISSA International board of directors, contributing author and founding member of the Cloud Security Alliance (CSA), and President, New York Metro CSA Chapter.

Fusco is certificated and accredited as a CISSP, CISM, CHS Level III (Certified in Homeland Security), National Security Agency INFOSEC Assessment Methodology Auditor (AIM Auditor), and National Cryptologic School Adjunct Faculty Certified Instructor (NSA/CSS/NCS) and holds an MS in Information Management.

Ilene Klein For almost 20 years, Ilene Klein has been evangelizing security to anybody who would listen … and to management. During that time, she built and led compliance, governance, incident response, privacy, policy, security awareness, and vulnerability management programs and frameworks. Ilene has earned multiple security and privacy certifications, and she won the CISM Geographic Excellence Award for earning the highest score in the North America geographical region on the December 2011 CISM examination.

Jeani Park is a leading security visionary who has held several executive leadership roles and created bridges to new horizons over the last 20+ years.

Over the course of Ms. Park's career, she developed products in several security areas: AV-anti-spam, network security, wireless security, client server security, mobile security, Internet security, systems management security, endpoint security, and identity management.

Ms. Park is a businesswoman and technologist who has enjoyed a 20+ year career developing, marketing, and monetizing technology products for organizations and consumers around the globe. She enjoys developing emerging technologies that solve real world problems across historically separated domain areas. Ms. Park has worked for companies such as QAD, Access360 (acq. IBM), HP, Dell, Mirage Networks (acq. TrustWave), and Veriato.

During the past fifteen years, Ms. Park focused on digital security, developing products for endpoint security, network security, forensics, user activity monitoring, and security policy and posture. In collaboration with world-class teams and technologists Ms. Park helped build: the integration of on-premise anti-spam functionality with gateway security products (Trend Micro); the integration of hardware and software Systems Management (at Dell); the Network Access Control space (at Mirage Networks); and the User Activity Monitoring space (at Veriato). Ms. Park also had the pleasure of serving as the Cybersecurity Director in the State of Maryland and is a published security author and speaker.

Recently Ms. Park has embarked upon Tech Venture Commercialization work with the University of Utah. Here, she mentors and encourages both tech and science entrepreneurs and researchers to realize success via commercially viable services, products, and companies.

Ms. Park is active in the community as a volunteer and donor for Make-A-Wish Utah, the Leonardo, the National Ability Center, the 1st Lego League,

Adopt-a-Native-Elder, the Humane Society of Utah, the Girl Scouts of America, etc. She is also a friend of the Library.

As a new Utah resident, Ms. Park enjoys the gorgeous scenery, myriad of outdoor activities, and looks forward to hiking, skiing, and paddle boarding. Ms. Park recently finished her first novel, a young adult techno-thriller entitled The Femme Fatales, Book 1.

M.L. Peck is the chief content and marketing officer at the Institution for Supply Management (ISM). ISM is the preeminent global provider of learning and professional development for supply management practitioners across the entire career spectrum. M.L. has nearly two decades of experience shaping corporate strategy in for-profit and not-for-profit organizations. Working closely with business and supply chain leaders, subject matter experts, volunteers, and business partners, the central focus of her role is to spearhead development of leading professional development programs and services for procurement and supply management professionals. Her zeal for analyzing market demands, technology trends, and the business landscape provides a fresh approach to strategic initiatives that ensure growth and profitability. She strives to bring out the full potential of those she oversees and avidly encourages an environment of continuous learning, coaching, and development.

Prior to joining ISM, M.L. worked at the Del E. Webb School of Construction at Arizona State University where she was instrumental in developing, promoting, and implementing three chief initiatives: a certificate program in construction management; a collaboration with Tecnológico de Monterrey in Mexico to deliver online programs to the U.S. and Mexican construction industries; and the first conference to bring together construction industry executives and Native American Tribal Leaders. She also held leadership positions at companies that develop software for the construction industry.

Adriana Sanford is currently a Visiting Professor at the Universidad de Talca in Chile and a Dean's Visiting Scholar with Georgetown University Law Center. She is an internationally recognized privacy and cybersecurity scholar and a leading expert on transatlantic relations.

Dr. Sanford is a Chilean-American author, professor, and international corporate lawyer who has worked with a broad range of global organizations to strengthen responsibility and transparency. Her specialization includes privacy, cybersecurity, international law, transparency, anti-corruption, and social responsibility. She lectures for numerous nonprofit organizations, regulary collaborates with the private sector, and presents at numerous events on global issues, including keynote addresses through the American Program Bureau.

Dr. Sanford will deliver the keynote address for the 2017 UTALCA/CIEPLAN seminar "Sociedad sin corrupción: el rol de las empresas y la politica," which includes distinguished speakers—Manuel Marfán Lewis, Alfredo Moreno Charme, and Alejandro Ferreiro Yazigi. In April 2017, she delivered the keynote address for the ISSA CISO Executive Forum conference on "Information Security, Privacy and Legal Collaboration," sponsored by Hewlett-Packard Company. In 2015, she delivered the keynote address for the Conferencia Latinoaméricana CACS/ISRM, the largest IT, telecom, technology, and security conference in Latin America for Spanish-speaking professionals specializing in audit, security, cyber ethics, cybersecurity, governance, and IT risk.

Dr. Sanford is also a regularly featured subject matter expert (SME) on CNN Dinero at CNN en Español, the network's 24-h Spanish language television broadcasting to more than 24 million viewers in the United States and throughout Latin America. Additionally she is the executive producer and host of "A Global Perspective with Dr. Adriana Sanford" with Manufacturing Talk Radio, sponsored by a global multinational risk management—Willis Towers Watson. Dr. Sanford interviews seasoned thought leaders and executives from some of America's leading corporations about theory and practice of data security, privacy, and safety in an interconnected world. Her senior correspondents include Wilfried Grommen, Chief Technologist Officer for Hewlett-Packard Enterprise (EMEA), Tawanda Mutasah, Senior Legal Director of Law and Policy for Amnesty International Secretariat, and Bruce Zagaris, author of "International White Collar Crime: Cases and Materials."

Dr. Sanford is the lead author of two business ethics books that provide critical insight for current and aspiring security, business, and supply chain professionals, *Business Ethics: A Guide to Surviving Storms, Challenges, and Ethical Risks* and *Ética Empresarial: Una perspectiva global.* She has also published a number of articles with the Institute for Supply Management that cover important and timely aspects of global corporate ethics and compliance.

Through education, outreach, speaking engagements, and the protections of rights, she continues to build awareness of, and advancement for, women in leadership as well as social justice and corporate responsibility. Dr. Sanford serves on the advisory committee of the World Economic Forum's Partnering Against Corruption Initiative (PACI), a leading international foundation and business voice on anti-corruption and transparency. Among her current research projects, she worked with the World Economic Forum on exploring levels of trust in the infrastructure and urban development industries. In 2016, Dr. Sanford worked on Building Foundations for Trust and Integrity, which is a new World Economic Forum report to help address corruption in Latin America. The report featured Mexico as a case study—as a potential regional model for reform.

Dr. Sanford is a member of the board of directors of Amnesty International USA, which is the largest country section of the world's largest grassroots human rights organization. She also chairs the advisory board for the International Institute of Management's think tank research; Mexico's Deputy Secretary of Education Javier

Treviño serves among the vice chairs of the think tank. Her current scholarly research project focuses on the challenges in managing intellectual property in the new age of Internet publishing.

Previously, she served as a delegate to the Asia-Pacific Economic Cooperation CEO Summit and as a Visiting Research Professor for UTALCA. For five years, she also taught law and ethics, ethical leadership, ethical issues of managers, and international management at W. P. Carey School of Business at Arizona State University to more than 1600 graduate and undergraduate students on an annual basis.

In 2015, she also served as ASU's Lincoln Professor of Global Corporate Compliance and Ethics to 85,000 students and faculty and as ASU's residential college faculty fellow for the College of Liberal Arts & Sciences. In 2011, she served on ASU's advisory board of the Ira A. Fulton Schools of Engineering EPICS Gold program, a national award-winning social entrepreneurship program designed to provide real-world opportunities for undergraduate students to change their world. Dr. Sanford's particular EPICS project focused on technology and sustainability to address social challenges in connection with 8.8 Richter scale earthquake that struck south-central Chile in February 2010.

Prior to teaching, Dr. Sanford served as the primary US counsel to several multinational businesses, including one of the largest international trading companies in the Southern Cone region—Argentina, Brazil, Chile, Paraguay, and Uruguay. Her clients also have included, among others, one of the world's leading security companies, one of Mexico's largest infrastructure companies, and one of Chile's largest fisheries. Among her in-house positions, she has served as the Assistant General Counsel of a trade finance bank, as well as lead counsel for Latin America and US commercial law counsel for four divisions of a Fortune 50 company.

Academically, Dr. Sanford completed 6 years of law school, receiving a Juris Doctor degree from the Notre Dame Law School and a dual Masters of Law (LL.M.) degree from Georgetown University Law Center in taxation and international and comparative law. She was the first Latin American woman accepted into Georgetown's dual LL.M. program and the only dual LL.M. in her graduating class. Dr. Sanford also studied international and comparative law in the year-long Concannon Program of International Law at the Notre Dame London Law Center in London, England. She then attended Thunderbird School of Global Management for postgraduate studies in global management. Dr. Sanford attended an all-girl's academy in Chile, Villa Maria Academy, before receiving her B.A. in political science from Arizona State University, where she focused on comparative politics. She is fluent in English, Spanish, Portuguese, and French.

Dr. Sanford is a 14th generation Chilean and a direct descendant of one of the oldest families of Cologne, Germany. She is a member of the Overstolz dynasty, which was the ancestral family of Cologne patricians. Her son attends California Institute of Technology and her daughter is in a rigorous high school STEM scholar diploma program

Evelyn de Souza is a security and privacy strategy leader currently developing architecture prototypes for IoT (Internet of Things) and cloud. She is on the Board of Advisors for the 360ofme, a data privacy startup, an Advisor to the Cloud Security Alliance, and a part-time consultant with Radius Consulting. She focuses the remainder of her time on an Affordable Housing initiative in her Coastside community. Evelyn has previously held leadership roles in the Office of the CTO (Chief Technology Officer) at Cisco and Intel Security. She-co invented the first business consumable data protection model and instigated OpenSource Privacy, a model for monetizing consumer privacy data. Evelyn was named as one of CloudNOW's Top 10 Women in Cloud Computing in 2014 and SVBJ's 100 Women of Influence in 2015.

Part I
Introduction

Chapter 1
Introduction - Book Summary

Debra Christofferson

Managing Cybersecurity Risk for the Coming Decade

We have come a long way Baby! What we have here is a showcase of women in the cybersecurity field, who have helped shape the face of security. Most are seasoned professionals with 20–30 years of experience. We have actively participated in growing this field from scratch. It has evolved from obscurity to present-day boardroom visibility. Sometimes, we have to take time to look back, to see just how far we've come.

This career choice has challenged us at most every turn, though it can also be rewarding. As one of the most in-demand career paths, most organizations are plagued by shortages in their talent supply line. We can help fill that gap.

Scope

Female security professionals are invited to share their own story of technology and innovation in security today, the foundation, where research is headed, and what the emerging trends are.

We also hope to increase the visibility of women in the field and their contributions. Women make up a small pocket of cybersecurity staffing. We can help fill the shortage of cybersecurity talent, and influence other women to join the field.

We are talking here to anyone who wants to seek greater opportunities in the field, college students seeking or completing a security or computer science degree, business leaders desiring a broader perspective or who want to increase female staffing, organizations serving cybersecurity professionals, and to those who create leading edge training programs for the cybersecurity workforce.

D. Christofferson (✉)

D.A. Christofferson (ed.), *Women in Security*, Women in Engineering and Science, https://doi.org/10.1007/978-3-319-57795-1_1

This book showcases a handful of players among many, and their roles in the field. We hope it increases your awareness of opportunities and trends in security, and encourages you to join the field.

Chapter Introduction by Author

M.L. Peck: "Preface: Women, Security, and the Supply Chain"

Security breaches have become commonplace. Third parties in the supply chain have contributed. We depend on cybersecurity. Technology continues to rapidly evolve, with the Internet of Things (IoT), cloud computing, wearable devices, and digital currency. Networks are ubiquitous where everything is computerized and connected. The digital economy and big data grow the attack surface, while hackers profit. Risks to businesses are enormous.

Tremendous opportunities exist as a result. Supply exceeds demand for these skilled positions. Today, women make up 10% of this workplace. We can grow our workplace presence beyond these staffing levels. Cybersecurity is a place where women work well, and we can make a greater impact.

Adriana Sanford: "Under a New Security Landscape: Global Ramifications of Multijurisdictional Conflicts"

Adriana presents a different view and supporting facts, on the global ramifications of privacy, security and technology, and cross-border data transactions. Greater scrutiny, accountability, and compliance dominate the global scene, and collaboration across governments and the private sector.

Dr. Sanford shares fresh insights on the Edward Snowden affair, invalidation of the US-EU Safe Harbor Agreement, privacy and data protection reform, legal access to offshore data, encryption's growing role, the impact of bulk data retention and the ongoing struggle with mass surveillance.

We have our work cut out for us, and you just might need a lawyer on staff to figure it out!

Debra Christofferson: "Security Risk Management, Women in Security, Resources and Certifications"

Ms. Christofferson presents security risk management from a tops-down view. Security risks garner boardroom attention today for most organizations. Cybersecurity is a business risk. Regulatory requirements often underpin security programs.

Security Magazine recently named regulatory changes as the C-Suite top enterprise risk. We advise you to look at the aggregate of risk, rather than just the immediate opportunity, and leverage points to best drive your security and compliance needs.

Technology evolves quickly and continually presents new risks. Risks include technology, people, processes, and other aspects of the business. A tradeoff always exists on the cost and lost opportunity tradeoffs to manage risk. Executives are responsible for the decision on how much risk the business takes. Our responsibility is to clearly articulate the risks and alternatives to support their decision-making.

Female-Specific Support Groups are covered briefly: We review some of the places outside of security for women to receive career support, and resources where you can further engage. A male perspective is offered, for women participating in female-specific support groups.

And finally, a short guide is given, on certification drivers, and which ones are most predominant for cybersecurity careers.

Diane Barrett: "Improving Women's Participation in the Security Field"

There are ways to be involved in security without a 9-5 job. Diane shares alternative career options for women, including professional writing in the field, and academia. You can become more involved through education and publication opportunities. This also helps you grow your knowledge and skills.

Publication also adds credibility and visibility.

This was how Dr. Barrett started her career, though education, certification and as a technical college instructor. She has written many security books. Academia and industry both require educational rigor.

Diane currently has shared roles in industry and education. She talks about why the number of women in the field is so low, and how to increase the number. Dr. Barrett hopes to directly influence those numbers. She shares her knowledge and showcases the profession through speaking, writing and teaching.

Everyone in the field for any length of time has their own network of contacts and connections. You should too. Diane includes resources for women in the field.

Evelyn DeSouza: "The Era of Homo Digitus"

We are experiencing a data deluge revolution, brought about by the convergence of the digital and physical world. While most technology has improved lives, data is also exploited and misused. Examples include data breaches, privacy infringements and identity and data theft. Be vigilant on how your data is shared and transmitted and about your device's privacy settings, and oversight of our children. Look at the lifespan of data collected in the cloud, and safeguarding your data—the lifeblood of the enterprise.

Technology is always transforming, but our cybersecurity approaches often lag as much as 10 years behind the business landscape. We should focus on less fear-based messaging, and more on technology as an enabler.

Ilene Klein: "This Girl Wants to Go to School Here!"

You need to know a lot of things to be a security leader, but you never know everything! It is a field where you are always learning, because the field is always changing.

Security standards and tools were very young when I entered the field. Fire drills were rampant—Microsoft patching and virus and malware outbreaks. We were at the forefront of security standards, as the BS7799 (British Standard) evolved into ISO27000 (International Standards Organization) and early vendor security standards pre-PCI (Payment Card Industry).

You need to be smart and knowledgeable to succeed in this field. Writing and speaking skills will serve you well, and an area where our field can improve. Degrees and certifications are expensive, time-consuming, difficult, and may not guarantee your knowledge level. But they often are "admission price" for a job.

Read Diane's chapter for writing opportunities. Debra can provide you with a quick guide to increase your credibility, visibility and skills in public speaking. (P.S.: All speaking is public speaking.)

Jeani Park: Three Decades of Digital Security

As a 30 year pioneer and innovator, Ms. Park shares advances in computing, big data, and where digital security is headed. Security has evolved across computer viruses, spam, patching and configuration, digital certificate authorities, fake sites, and mobile computing. Security is always a moving target, where the hackers parry, reinvent and adapt.

We can do better! From the Wild West of the past, we are moving to a new generation of security. Advances in computing and manipulation of big data will involve data analytics, deep learning and artificial intelligence. Metrics will drive risk and reward, and we will enjoy healthy competition and choice in the security marketplace.

Juanita Agard: The Pursuit of a Cybersecurity Career

If you are seeking a security and technology career path as you head into college, read on.

Juanita maps out a perfect strategy for education and pursuit of an IT career that morphs quickly into cybersecurity. Upon receiving her undergraduate degree, she had an IT consulting job waiting upon graduation with IBM. After later moving to Booz Allen Hamilton, Ms. Agard enrolled in a master's degree program which is near completion.

Juanita exhibits drive and an ability to set and complete goals, while keeping her eye on the prize. She is a full time parent, full time IT consultant, and a student completing her master's degree. Ms. Agard holds an active security clearance—always valuable. She earned the CompTIA Security + and the CEH—Certified Ethical Hacker certifications.

As you read other women's achievements here, you will see that Juanita has followed a precise recipe to land in the field: An engineering and math focus in her undergraduate degree, a cybersecurity master's degree (completing this year), certifications, continuous learning, and working for large consulting organizations. Ms. Agard's next goal is once she graduates is to pursue a law degree in technology. Her path is clear ahead.

Lacey Chong: "Making it in the National Security Field as a Millennial Minority"

It can be challenging to work as a young minority in the US Defense Sector in cybersecurity, where little diversity exists.

In Hawaii, my career began as a political-military Intelligence Analyst working for the U.S. Pacific Command at Pearl Harbor, before I moved on to other U.S. government organizations. My academic background—a bachelor's and master's degree—are non-traditional in the cybersecurity field.

More recently, I transitioned away from being a cybersecurity and counterterrorism expert. Instead, I moved towards business process improvement and project management.

The work is essential since threats against the United States are always evolving, and heightened by technology and social media. The restrictions on information sharing make timely relevant communication difficult between intelligence agencies. Groupthink and a lack of diverse ideas are problems in the national security field.

Soft skills can positively impact your ability to get things done. When making change, the people and communication aspect are often left out.

Martha Daniel: "Hidden Dangers of the Internet of Things"

We have progressed in computer security to wireless telecommunication to integrated data and connecting globally over the Internet. It has changed the way we do business, live and work.

New avenues have opened today. We are subject to global rules and regulations, collaboration, and protection requirements across countries. There are always those who wish to capitalize financially through theft, corruption and fraud. But the benefits of innovation far outweigh the negatives. However, the ones who wish to do us harm have grown significantly in their presence. The challenge is in keeping the bad people out.

The number of Internet connected devices worldwide—the Internet of Things—IoT is growing exponentially. That is what this chapter is about, because IoT is so dominant in growth. IoT includes laptops, tablets, smartphones, smart home devices for lights, climate, security, and household appliances, Smart TVs, automated door locks, remote garage door openers, baby monitors, video cameras, and connected automobiles…for starters. They have differing capabilities. Safety, healthcare, automation, inventory control, asset tracking, traffic control, and automation are a few of the benefits.

However, many IoT devices lack security and any consideration of the data and network capability in place during design and implementation. This creates global risk. There are solutions of course, and like all new technology, we are working to increase cybersecurity and manage the risks.

Mary Ann Davidson: "Information Security: Beyond the Bits and Bytes"

Security is a hard job. You have to be adaptable and ready for change at any time. Most of my professional career has been in information security.

Security is bigger than just technology. You have to look at the capability, the benefits it brings, and what unintended consequences could occur. Information security today has spread to "networking everything". Moving to cloud computing does not diminish the importance of security.

When building a security team, you want really good people with a variety of skills, and a variety of viewpoints—with the freedom to express them. Hard skills and soft skills are both important.

Filling the cybersecurity hiring gap goes beyond just hiring more people. It's a problem of engineering our systems to be more secure and more attack resistant.

Cybersecurity is a very challenging field. Your work environment will be a key factor in your success.

Miriam Fernandez: "Building the Bridges to Security"

Security is an evolution that happens over time. It is not based on a single action or accomplishment. Security shifts and changes constantly, along with your own career within. Sometimes it is hard to see the light at the end of the tunnel. Work is often progressive and longer term before you see results.

In Miriam's career, the network evolved from raw cable to more complex networks, and to the wireless, mobile, hosted and the cloud environment you see today. Changes that she helped execute in her IT roles: Standardizing server builds, automating changes, remote trouble-shooting, migrating to a virtual server structure to support offshore outsourcing decisions, and creating disaster recovery processes. Collaboration is very important to getting things done across teams and groups.

IT engineering and operations present a strong starting or transition point for those wanting to contribute to security innovation. Continuous improvement and ongoing growth are requirements to survive, and an IT degree is needed to land most any role today. The field moves fast. You are always learning on the job. Many career opportunities exist, where you can also make a difference for the future.

Pamela Fusco: "It's All About the Cocoa Beans"

Ms. Fusco got her start in the Navy. She was a cryptographer, and has given herself a "Cyberologist" moniker. Ms. Fusco is an experienced executive leader and former CISO in several large organizations. Her chapter will talk about the military value of building trust and relationships early on in her career, while working in different places and times with people she did not necessarily know. Read her chapter for her story on why the cocoa beans matter so much.

Pamela was a pioneer in digital e-commerce who helped shape hosted services as we see them today. She was also instrumental in creating better solutions in managing malware outages. Her company was not taken down by the early ravages of business paralyzing worms and viruses. She designs for the 80/20 rule when it comes to the technical infrastructure and the customer base it supports. A finite set of builds are offered for a customer to choose, which greatly streamlines their change management process and ability to respond to potential security incidents. Ms. Fusco's organization also shared industry leading practices with Microsoft for a revised customer patching process.

Pamela Fusco will share insights on what it takes to enter and succeed in the field today, and predict where the field is headed—towards her next mission.

Rhonda Farrell: "Organizational Intelligence—Cybersecurity as a Performance Optimizer"

Dr. Farrell presents a Baldridge model from NIST—the National Institute of Standards and Technology for business intelligence and process improvement. It is a continuous improvement program to increase the organization's performance, personnel optimization, innovation, and GRC—governance, risk, and compliance management. Cyber generated data is examined through a business intelligence lens. One key is to always ask "why" something is being done, "how", and "what the effect is" on the organization, products, services, stakeholders and workforce.

Rhonda is heavily involved in the industry beyond her full time associate role with Booz Allen Hamilton in the US Government's Defense and Federal sectors. She supports IEEE heavily in committee leadership for Northern Virginia and Silicon Valley. Rhonda founded ISSA's Women in Security Special Interest Group (WIS SIG).

Conclusions

New technology is nothing new. When addressing cybersecurity in the business, there are always risk trade-offs in creating business opportunity choices. Criminals follow the money and today that means digital currency and online access. Focus on leverage opportunities, and high risk and high return in your security program.

Seek innovation opportunities—where you see overlap, redundancy and gaps. Look ahead at where the market is headed and get ahead of the curve. Within your own organization or community, watch for the places and ways you can make a difference.

Some of the younger generation can feel alone and shut out as a minority in the field. We can be more open to different viewpoints and ways of thinking—it can help us innovate. Also, increasing diversity and the number of females in our field is good for women, business and the economy. When they become more visible, the numbers increase from good hiring, and from greater retention.

The greatest demand is for security analysts—the technical professionals supporting the infrastructure. Entrepreneurs are always in vogue, and many people make big money by starting up security technology and services companies.

Build new skills for the new marketplace—whether you are a hiring manager responsible for staff development, or as anyone plugged into the workplace.

Build your network and connections. Become actively engaged in the industry—it will reap big rewards to your career at every stage.

Part II
Cyber Risk, Technology & Innovation

Chapter 2
Under a New Security Landscape: Global Ramifications of Multijurisdictional Conflicts

A Guide for CEOs and Board of Directors on Multijurisdictional Legal Issues

Adriana Sanford

The new security landscape requires a deeper level of scrutiny, greater compliance, increased accountability, and more collaboration among our governments and the private sector. Encryption failures, the emergence of sophisticated hackers, and vast data collection by data brokers in our post-Snowden world exposed the complex correlation between security and privacy.

As regional security remains a concern for both the United States and European Union with the disturbing trend of terror attacks, governments are scrambling to protect their national interests in an inter-connected world—often at the expense of privacy and perhaps other individual freedoms. In a world that has become more commercially and professionally connected by modern technologies, drones flying overhead, apps tracking every move, and personal data flowing into cyberspace every second, the struggle to safeguard citizens will continue to intensify.

Snowden Revelations

Risks of harm to privacy interests in the EU escalated after former US Central Intelligence Agency employee, Edward Snowden, captured the attention of viewers in 2013 with revelations of extensive US and British mass surveillance. Snowden revealed to the media an extensive internet and phone surveillance scheme, which included the collection of the telephone records of tens of millions of Americans

---

STEM scholar diploma student Sofia A. Schmidt-Koeck assisted with the preparation of this article.

A. Sanford, B.A., J.D., Dual LL.M.in Taxation and International & Comparative Law (✉)

D.A. Christofferson (ed.), *Women in Security*, Women in Engineering and Science, https://doi.org/10.1007/978-3-319-57795-1_2

and the storage of approximately 200 million text messages per day across the globe.[1] The surveillance operation spearheaded by US intelligence had the ability to read almost all smartphone information, including SMS, location, emails, and notes.

The National Security Agency's (NSA) Prism data collection program targeted France, Italy and Greece, as well as America's non-European allies such as Japan, South Korea, India and certain regions of Latin America. The NSA used the assistance of the British Government Communications Headquarters and the servers of nine internet firms to track online communication, which included Facebook, Google, Microsoft and Yahoo. The Snowden revelations made it clear that the government was asking tech companies for private user information. As a result, boundaries between commercial and government data were, at best, blurred.

EU complaints under regional data protection law were directed against Facebook, Apple, Microsoft, Skype and Yahoo for their alleged collaboration with the surveillance program. The key issue was whether "mass transfer" of personal data to a foreign intelligence agency was legal under European law, as the European subsidiaries of these five US companies were clearly subject to European privacy laws. Because of their corporate structure, despite being headquartered in the US and required to comply with US surveillance requirements, the companies were also required to comply with EU legal requirements, which were in conflict with one another.

Invalidation of US-EU Safe Harbor Agreement

The EU determined that the mining of data from consumer web services breached the EU data protections laws. Not surprisingly, this led the European Court of Justice (ECJ) to invalidate the longstanding US-EU Safe Harbor Agreement in October 2015, ending a mechanism that had authorized personal data transfers between the EU and the US for fifteen years. The European Court of Justice held in the *Maximillian Schrems v. Data Protection Commissioner Case* that the Safe Harbor Agreement failed to provide sufficient safeguards to protect EU citizens' personal identifiable information.

The landmark privacy ruling, which affected the ability of more than 4000 US companies to transfer and collect personal data from EU countries, noted that the EU Commission must take into account US mass surveillance practices and surveillance laws, as any data that US companies collect could potentially be available to US intelligence.

At the core of the decision was the belief in a fundamental human right to privacy, which is strongly supported in the international community and memorialized in the UN Declaration of Human Rights.

[1] Mardell, Mark. "Report: NSA 'collected 200m Texts per Day'." BBC News. BBC, 17 Jan. 2014. Web. 25 Jan. 2017. http://www.bbc.com/news/world-us-canada-25770313.

The EU has one overarching law that regulates privacy consistently across all industries. In contrast to this omnibus approach, the US, as a common-law country that evolved a multidimensional system of federal and state laws, has a sectoral approach where laws are directed to specific industries. The most sensitive data—such as financial, medical, health, electronic communications and children's information—are protected in the US by nearly two dozen federal sector-specific laws and numerous state laws, which results in a web of hundreds of statutes, along with regulations and sometimes binding industry standards.

Over the last three years, US tech executives have expressed significant concern to law makers and multiple regulators about competing regulations, conflicts of laws, and unintended consequences relating to the access to personal data stored outside the reach of the US warrant authority as well as US proposals to weaken encryption.

EU-US Privacy Shield Framework

In July 2016, European Commission and the U.S. Department of Commerce jointly adopted the EU-US Privacy Shield framework, which replaced the invalidated Safe Harbor agreement and ended the nine months of uncertainty. In the meantime, US companies relied on more cumbersome legal mechanisms in the wake of the October 2015 ECJ ruling.

The new mechanism was designed to bridge two distinct legal regimes with an aim to achieve 'essential equivalence' of European data protection laws in the US and provide some legal certainty for businesses operating in the two regions without substantial reform of US laws. Approximately 1500 companies are currently using the new framework.

The Privacy Shield provides EU citizens with a means to seek redress in the event of US spying, including a new privacy ombudsman within the US State Department. While the EU Commission believes that the Privacy Shield will live up to the requirements set forth by the ECJ, a legal challenge has been filed by Digital Rights Ireland. At this stage, it will be a year or more before the court rules on this case. The court could also declare the case inadmissible if the court finds that the Privacy Shield is not of direct concern to this Irish privacy advocacy group.

It remains unclear, however, what impact the US Executive Order (Enhancing Public Safety in the Interior of the United States), along with US President Donald Trump's intent to revise the US privacy regulations, may have on the Privacy Shield arrangement. The European Commission issued a statement indicating that the current US Privacy Act has never offered data protection rights to European citizens. Accordingly, "two additional instruments [were negotiated] to ensure that EU citizens' data is duly protected when transferred to the U.S."[2]

---

[2]DiPietro, Ben. "The Morning Risk Report: Trump Executive Order Jeopardizes U.S.-EU Data Pact." The Wall Street Journal. Dow Jones & Company, 27 Jan. 2017. Web. 31 Jan. 2017.

Data Protection Reform

The EU data protection regime, namely the EU Data Protection Directive 95/46, was widely considered to be inadequate in light of the Snowden disclosure and the revelations caused considerable political outrage in most of Europe. Reform was needed to protect the rights of citizens. The General Data Protection Regulation (GDPR), which enters into application in May 2018, simplifies the regulatory environment for business, while banning the transfer of data to the US or third countries unless based on EU law or under a transatlantic pact. Specifically, the GDPR harmonizes data protection law across the European Union and establishes a 'one-stop-shop' for businesses. Under the current EU Data Protection Directive, each EU member state enacts their own set of rules, which creates a patchwork of data protection regimes throughout Europe that sometimes conflict with each other. Although the GDPR simplifies the process and makes it cheaper for US companies to do business in the EU, there are heftier fines for non-compliance that may include up to 2% of global annual turnover.

Tech Industry and Executives Under Pressure

Progress in technologies coupled with globalization have profoundly changed the way personal data is collected, accessed, and used. Legal debates concerning personal data, encrypted communications, and mass government surveillance continue to mount. While government officials across the Atlantic and law enforcement struggle to defend citizens from real security threats, technology companies urge governments to endorse consistent principles and enact reforms that will protect cybersecurity, economic growth, and human rights.

Access to Offshore Data

A US court ruling upholding a US warrant which required Microsoft to abide by a US government request for personal data that was processed outside the US created a dilemma for the cloud computing industry. Consumer cloud service providers worried that such requests would hinder their ability to do business and store data in centers in Europe and other locations. Microsoft argued the US government needed to go through a procedure outlined in the mutual legal-assistance treaty (MLAT) between the US and Ireland.

In December 2013, a New York district court judge issued a US warrant instructing Microsoft to produce all emails and private information associated with a certain account hosted by the company and stored on a server located in Dublin, Ireland. Microsoft refused to turn the emails over and responded that a US judge had no jurisdiction to issue a US warrant for information stored abroad. In December 2014, the Irish government filed an amicus brief in support of Microsoft maintaining that the emails should be disclosed only on request to the Irish government pursuant to the long-standing MLAT.

Although a three-judge panel of the Second Circuit ruled unanimously in favor of Microsoft, on October 13, 2016, the US government filed a petition for rehearing with the Second Circuit, as they were concerned that companies will deliberately offshore their data storage in order to prevent access to US law enforcement with SCA-citing warrants.[3]

The data in these cases could potentially be out of reach for all jurisdictions. US law enforcement would be unable to access the data because it is outside the reach of the US warrant authority. The data would also be outside the reach of an MLAT request since only US-based employees can access customer email accounts (regardless of where they are stored) and foreign governments lack jurisdiction over US-based employees that control the data. Consequently, no law enforcement official anywhere would potentially be able to access the personal information.

Perhaps one of the most serious challenges for senior executives is the concern of personal liability for non-compliance. In May 2016, Facebook's vice president for Latin America Diego Dzodan faced criminal liability in Brazil for Facebook subsidiary WhatsApp's alleged non-compliance with court orders. WhatsApp had indicated it was unable to provide the information requested by the Brazilian government about a WhatsApp user relating to an organized crime and drug-trafficking investigation, because WhatsApp messaging service does not store content. WhatsApp stated that it had "cooperated to the full extent of [its] ability in this case and while [it] respects the important job of law enforcement, [it] strongly disagrees with its decision."[4] Facebook also issued a statement, noting that the two companies were operationally separate from one another and the arrest of a Facebook executive was "extreme and disproportionate."

Encryption Explosion

In May 2015, Apple and 140 tech companies signed a letter urging the US government to preserve strong encryption against pressure from law enforcement and surveillance agencies. A sensitive privacy issue for many users, particularly for after the Snowden revelations.

---

[3] The Stored Communications Act (SCA) is the US law regarding voluntary and compelled disclosure of "stored wire and electronic communications and transactional records" held by third-party internet service providers.

[4] Meyer, David. "Brazil Arrests Senior Facebook Exec Over WhatsApp Aid In Drug Case." Fortune.com. Fortune, 01 Mar. 2016. Web. 26 Jan. 2017. http://fortune.com/2016/03/01/brazil-facebook-arrest/.

In part, the letter requested that policies be developed to "promote rather than undermine the wide adoption of strong encryption technology. Such policies will in turn help to promote and protect cybersecurity, economic growth, and human rights, both here and abroad."[5]

During 2015 and 2016, Apple objected to or challenged several orders issued by US district courts to compel Apple "to use its existing capabilities" to extract data from locked iPhones in order to assist with criminal investigations. The February 2016 legal tussle between Apple and the US Federal Bureau of Investigation (FBI) marked another high-profile clash. The FBI requested that Apple write new software for an iPhone to bypass security and allow the FBI access to information about one of the shooters of the San Bernardino attack. When Apple declined to create the software, a US magistrate judge issued a court order mandating that Apple create and provide the requested software. Apple's CEO, Tim Cook, stated that the legal order "had implications far beyond the legal case at hand".[6]

In March 2016, the FBI revealed that a third party managed to unlock the iPhone and they withdrew their request. The underlying controversy and possible outcomes over encryption and data privacy are problematic. The use of the All Writs Act to compel Apple to write new software was unprecedented and indicates the importance of striking the right balance between privacy and security.

Not surprisingly, in April 2016, Apple, Microsoft, Google, and Amazon further expressed their concerns over a controversial encryption bill that had a mandatory decryption requirement. This requirement would compel smartphone makers to decrypt data on demand. In part, the companies argued that "the effect of such a requirement will force companies to prioritize government access over other considerations, including digital security".[7] Also noted was that the "no accessibility requirement" could not be limited to US law enforcement and that other governments (or even hackers) could potentially follow. Tech companies asserted that encryption was a necessary step, particularly with the growing proliferation of smartphone banking.

They argued that encryption had the ability to protect us from innumerable criminal and national security threats, including the common street criminals, online computer criminals, corporate spies, repressive governments, or foreign intelligence agencies trying to compromise sensitive national security secrets.

[5] Goldman, David. "Apple and Google to Obama: Hands off Our Phones!" CNNMoney. Cable News Network, 19 May 2015. Web. 25 Jan. 2017. http://money.cnn.com/2015/05/19/technology/apple-google-obama-phone-encryption/.

[6] Srivastava, Shivam. "Apple CEO Opposes Court Order to Help FBI Unlock IPhone." Reuters. Thomson Reuters, 17 Feb. 2016. Web. 25 Jan. 2017. http://www.reuters.com/article/us-california-shooting-timcook-idUSKCN0VQ0XW.

[7] Alba, Alejandro. "Tech Giants Publish Letter Expressing Concerns over U.S. Bill." NY Daily News. N.p., 20 Apr. 2016. Web. 25 Jan. 2017. http://www.nydailynews.com/news/national/tech-giants-publish-letter-expressing-concerns-u-s-bill-article-1.2608292.

Search Warrants and Gag Orders

Microsoft's lawsuit against the US Department of Justice (DOJ) in April 2016 set the stage for another high-profile confrontation between the government and a tech giant. The company challenged portions of the 30-year-old Electronic Communications Privacy Act (ECPA), that permitted the government to secretly access information in the cloud computing era with ECPA provisions enacted prior the rise of the commercial Internet and the emergence of cloud computing.

According to the lawsuit, the government is violating the US Constitution by serving the search warrants on service providers without the user's knowledge, which contravenes the Fourth Amendment. According to Microsoft, the US government used the transition to cloud computing as a means to conduct secret electronic investigations. In 18 months, Microsoft received 5624 federal demands for customer information. Nearly half of the requests contained gag orders that prevented Microsoft from informing customers, which according to the lawsuit violates Microsoft's First Amendment right to Freedom of Speech. Moreover, some of the secrecy orders had no time limit, which indefinitely gagged the company from telling its customers that the government obtained their digital files.

In January 2016, a settlement was reached with Google, Yahoo, Microsoft, LinkedIn and Facebook on their ability to disclose surveillance orders. After those firms withdrew their legal challenge, Twitter attorneys met with the US Justice Department and FBI. Twitter argued that they did not receive the same scale of surveillance requests as the five companies and therefore should not be subject to the limits set forth in the settlement, but the US government refused to amend the agreement. Twitter, the world's largest microblogging platform, then filed a lawsuit against the DOJ.

The lawsuit is part of a broader push for surveillance reform through legislation, such as the USA Freedom Act pending in the Senate. Twitter alleged that the US government's restrictions on what the company could disclose publicly about the national security requests for user data violated its First Amendment rights of Free Speech. In the complaint, Twitter stated that the government's position "forces Twitter either to engage in speech that has been preapproved by government officials or else to refrain from speaking altogether."[8] Several organizations, ranging from large tech companies to news organizations and others, have filed legal briefs in support of Twitter's legal case.

---

[8] Nakashima, Ellen. "Twitter Sues U.S. Government over Limits on Ability to Disclose Surveillance Orders." The Washington Post. WP Company, 7 Oct. 2014. Web. 25 Jan. 2017. https://www.washingtonpost.com/world/national-security/twitter-sues-us-government-over-limits-on-ability-to-disclose-surveillance-orders/2014/10/07/5cc39ba0-4dd4-11e4-babe-e91da079cb8a_story.html?utm_term=.fe335a7cdbd4.

Bulk Data Retention

Bulk data retention is the blanket storage of all communications data passing through the networks of communications providers. ECJ has declared that the "general and indiscriminate retention" of personal identifiable data is inconsistent with privacy rights. In December 2016, the ECJ set new precedent for all EU member states on any data retention procedures. The Luxembourg-based court ruled against the bulk retention of emails and other electronic data by EU member states and that any such "national legislation exceeds the limits of what is strictly necessary and cannot be considered to be justified within a democratic society".[9]

This ruling comes after the adoption of the Investigatory Powers Act on November 2016, a British legislation that granted new bulk surveillance powers to police and intelligence services. This Act, among other measures, requires websites to keep customers' browsing history for up to a year and allows law enforcement agencies to access the browsing history to help with their investigations. Referred to as the Snooper's Charter, the Act also permits the UK government to legally track millions of devices, including computers and smartphones, as well as 'snoop' people's browsing history. Under the Act, companies are legally required to assist with surveillance operations, including instructions to bypass encryption. According to the ECJ, however, only targeted interception of traffic and location data is justified in order to combat serious crime.

On-Going Struggle with Mass Surveillance

In January 2016, the European Court of Human Rights (ECHR) ruled that the Hungarian government violated the right to privacy provided under the European Convention on Human Rights when the government failed to include "sufficiently precise, effective and comprehensive measures that would limit surveillance to only people it suspected of crimes".[10] ECHR referenced the December 2015 legal decision where the court held that the Russian government had also violated privacy rights with their mass surveillance of telephone calls.

Understandably, while surveillance demands rise sharply, tensions between governments and technology companies will continue to increase. Mandatory data retention or backdoors for encrypted communication are considered by some governments to be vital for modern investigative techniques and necessary to combat

[9] Behles, Caitlin. "Court of Justice of the European Union Rules States May Not Impose General Obligations on Data Retention (December 21, 2016)." American Society of International Law. N.p., 13 Jan. 2017. Web. 25 Jan. 2017. https://www.asil.org/blogs/court-justice-european-union-rules-states-may-not-impose-general-obligations-data-retention.

[10] "European Court Rules Mass Surveillance Infringes Basic Human Rights." Lawyer Herald. N.p., 24 Jan. 2016. Web. 25 Jan. 2017. http://www.lawyerherald.com/articles/29083/20160124/european-court-rules-mass-surveillance-infringes-basic-human-rights.htm.

organized crime and terrorism. Other countries strongly disagree, believing that it is only acceptable for authorities to engage in the "targeted retention" of data in cases involving serious crime. According to Microsoft's chief legal officer Brad Smith, people are "placed in a situation where [they] have to decide what law to break. It isn't a comfortable place to be" as companies face impossible and conflicting demands from multiple regulators in different countries.[11] Where the legislation of one jurisdiction competes or conflicts with the laws of another country, it is incumbent upon the governments to work together to resolve the underlying controversy.

Conclusion

Governments face an increasingly complex security landscape and can only begin to envision the technological challenges of tomorrow. Snowden's revelations opened the international debate on the *methods* for accomplishing surveillance. In an era that demands greater transparency and accountability, governments seem to find themselves pulled in opposing directions on whether mass surveillance operations are effective and justifiable. Some authorities believe that that these modern investigative procedures are a necessary and valuable source of intelligence to stop terrorist plots. However, others consider privacy to be a fundamental human right and mass surveillance, as well as any indiscriminant data collection, retention, or use, is viewed as a potential violation of such right.

Multi-jurisdictional legal predicaments have increasingly become a focus of attention for the tech industry, particularly C-suite leaders and their board members as governments continue to demand cooperation.

Significant progress has been made through the sharing of best practices and lessons learned, but "consistent" legislation in geopolitical areas is needed, as well as further training on many of the complex multijurisdictional challenges that currenlty exist. The private sector needs clear guidance as to the way forward through competing and conflicting requirements that can create personal criminal liability for executives.

Lastly, the conflicts involving offshore data access, encryption demands, electronic gag orders, and bulk data retention are merely some of the current issues involving privacy and security that warrant focus and attention. Matters that raise widespread privacy and security concerns also stem from the business-related or "commercial use" of certain evolving technologies. Without appropriate legislation, it is possible that advances in facial recognition technologies may ultimately end the ability to remain anonymous in public. Serious privacy and physical safety concerns arise when previously anonymous individuals on the street (or in a restaurant or bar) can be identified, in real-time, by a mobile app.

---

[11] Schechner, Sam. "Tech Companies Bring Battle Over Data to Davos." The Wall Street Journal. Dow Jones & Company, 20 Jan. 2016. Web. 25 Jan. 2017. http://www.wsj.com/articles/u-s-tech-companies-bring-encryption-battle-to-davos-1453320950.

As governments continue to amend and reform their data privacy and surveillance laws, it is expected that more conflicts and criminal exposure will emerge. It is anticipated, however, that GDPR will provide a basic framework for data privacy legislation to several other regions and countries around the globe. Complex jurisdictional and security issues must be addressed through a broader lens, because the solutions may be beyond one country's national interests and decisions may have strong repercussions on a global scale.

Chapter 3
Managing Cybersecurity Risk for the Coming Decade

Debra Christofferson

Opening

Security has become a boardroom issue today—no business can afford to ignore the risk. Breaches have become mainstream, criminal enterprises operate with impunity, sophisticated and destructive malware and denial of service attacks are commonplace, ransomware and cyber extortion are becoming mainstream, mobile business platforms present prime targets, and technology continues to outpace our ability to keep up. Cloud computing and the Internet of Things represent emerging technology risk.

Convenience nearly always trumps security and security solutions always lag. How much security is too much? What does a successful security risk strategy involve? What are the current and future trends impacting security and IT risk to your bottom line? What skills are required to evolve our future generation workforce? Where are the rising career opportunities? Join me in looking at challenges and solutions for the next generation of security risk and its impact on your workplace:

- Core concepts of information security and risk management
- Principles of risk management
- Where IT and security fit within enterprise risk management
- How to manage risk
- What a successful security risk strategy involves
- Current and future trends impact security and IT risk to your organization's bottom line
- Rising career opportunities

D. Christofferson (✉)

D.A. Christofferson (ed.), *Women in Security*, Women in Engineering and Science, https://doi.org/10.1007/978-3-319-57795-1_3

Defining Risk

Risk is a likelihood that damage will happen.

At the highest level, ask yourself these questions to create an immediate baseline:

What is the primary mission of your organization and how does security impact it?

How much can your organization afford to lose if some event occurs?

For instance, if you look at disaster recovery planning or business continuity: What would it take for your business to remain viable?

What steps could you take to avoid outages and disruptions that would impact that?

If certain events occurred, what would be the impact to your bottom line? Would your organization be able to stay in business and be successful?

Those risks must then be addressed to prevent failure.

Security decisions should be strategic and based on risk. Much of cybersecurity is based on technology as an enabler. But the function actually is based on your specific business risks, not technology or IT.

Your risk plan's purpose is to integrate security risk management into the company's enterprise risk management plan. If no central program exists, then your plan sets the foundation for your security program.

Your goal is to protect the company and its assets, by identifying, prioritizing and managing security risk to an acceptable level by:

- Identifying the company's security risk profile
- Keeping management informed of security risk to help them make more informed decisions to protect the business
- Aligning the plan to the business
- Creating a common definition and plan of action for identifying, managing and reporting on security risk

The security risk management plan is owned by the overarching security manager for the organization. Risk is owned by the most senior executives of the company, who also make the final risk decision.

Business Case for Risk Management

Your purpose in managing security risk is:

- Keeping the business viable
- Keeping the systems up and running
- Protecting intellectual property/information assets
- Maintaining revenue stream
- Preventing financial theft or fraud
- Keeping market share

- Meeting regulatory requirements
- Safeguarding reputation (keeping out of the news headlines!)

Business Drivers: WIIFM (What's in it for Me?)

WIIFM applies to the business stakeholders, and what they need—not you, and not security.

The business cares about the business, and security exists to support it:

- CEO accountability to control financial activities
- The need to manage risk
- Technology runs the business
- Duty to protect customers, shareholders, and communities

A company is in business to be profitable. Unless you are selling security solutions, security is not the primary business. Business viability does depend on a good security program. Non-profits and government organizations identify business drivers for their core business function.

Risk management is about balancing costs of a security failure to the business, against what it costs to prevent them. Security's mission is all about risk, and all security decisions should be based on business risk.

Common Denominators in Managing Cybersecurity Risk

In most all organizations—public, private, government, financial, healthcare, high tech, and others, these common denominators work:

Use a security framework: ISO is common in the private sector—the 27002. NIST is used typically in the government space, because it's a federal requirement. ISO is international, while NIST is specific to the U.S. They are easily interchangeable. My preference is the ISO standard, which created the foundation for distributed security, based on an early British standard, the BS7799 in 1995. ISO's model is easier, and many vendors apply it to certify their systems as secure. NIST is free, while ISO costs real money. Choose your model based on your business needs. Other models exist, but these are the two I would look at.

Create a security steering committee with a friendly senior stakeholder as Godparent and management sponsor. Include key stakeholders from IT, Legal, Audit, HR, and key business groups.

Automate security in the systems and technology infrastructure where possible.

Identify the core business.

Collaborate with the business to understand the primary mission and the security risks that impact it.

Avoid using compliance as your primary driver for funding and support. Yes, we do have to comply with regulatory mandates, and compliance *is* a risk. Even if you were hired primarily to support compliance, do not use it as your main driver.

Factor in the behavior elements inside and outside your organization, and processes that support security risk management. Do not rely solely on technology.

Brief your senior managers on risk in business terms. Avoid all technical jargon. Make risk a conscious decision, rather than a default event.

Manage the accounts—who has access to what, and audit them regularly for terminated and transferred employees, unused accounts, generic accounts, vendor accounts, privileged accounts, default accounts and passwords, and network devices.

Include security in the software/hardware and product life cycle.

Create a policy feedback mechanism for employees and stakeholders; review policies annually for reoccurring issues.

Creating a Successful Security Risk Strategy

First steps:

- Identify your core business
- List the business drivers
- Establish how security interacts with these business drivers
- Look at the risks in your specific environment

You can approach this a dozen different ways. Details will vary based on your organization's particular business, and their maturity and support for security.

To create your strategy at a high level:

- Look at the audit findings current and recent past.
- Interview business leaders and senior management on what they see as risks.
- Create a strawman—an initial rev.0 draft of your plan: Write down your initial thoughts based on what you found in the steps above and add what you see and know. (Ask me if you want an example copy of a security strategy framework.)
- Go out and collaborate with stakeholders in the organization, to verify and prioritize the outcomes.
- Take a cut at prioritizing risks at a high level, based on feedback from your management and the business leaders.
- Since you will not be able to do everything, you will then start on your roadmap, to implement solutions to these risks.

Risk Worksheet for Risk Management Planning:

1. Business Mission:
2. Enterprise risks that would impede this mission:
3. How your security charter supports the business mission:
4. What your security organization is chartered to safeguard:

5. Your organization's top 3 security risks:
6. How they relate to the business enterprise risk:
7. What about consistent or persistent audit findings?
8. Key Stakeholders:
9. Key Stakeholder who is a strong proponent of security:
10. Top 3 risks you can begin to address immediately:
11. Longer term risk management strategy to add to your business plan:

That is the start of risk management planning, before you look for a framework at the most strategic level.

Many people jump in with two feet before they get this far. They bring in multiple models, talk about them a lot, and then start filling in a model without a foundation in place. Risk models belong in the execution stage, not when you are trying to build a risk plan.

Another non-starter is to jump straight into risk "assessment". Risk assessment may be part of your implementation, but it is not executed during the security strategy phase. Without a risk plan and a security strategy to support it, the process will be analyzed to death without significant results and the effort will fail.

One Size Does Not Fit All

You can buy a risk strategy—and plenty of organizations are selling it. At a strategic level, one size might work as a starting place. You may be able to bucket organizations by sector and size, for some set of options. When you start to define what it means to your own organization, approaches will differ, based on sector, size, your own risks, and maturity of the program.

Global Business Risk Example

The best example of global business risk I've ever seen, which still exists today:

Somali piracy of shipping vessels along the Gulf of Aden on Africa's eastern coast was being reported in 2009, in the Indian Ocean. A cruise ship successfully fought off an attack with high pressure water hoses. A luxury yacht was taken hostage with vacationing tourists. Cargo ships were primarily targeted. They were boarded and taken over by pirates until a large ransom was paid for release. The armed gangs were highly organized and skilled.

Locals used this as their economic business model to support their villages—those taking the greatest risk got paid the most. Initially the percent of pirated ships was low. The world and the ship owners took greater notice and action as attacks and costs mounted, and violence burgeoned.

Why did the ship owners continue to navigate the vessels along this coast when they know about the piracy risk—vessel seizure, kidnapped crews, and ransom demands? While this is not a cybersecurity question, but it definitely creates physical security and supply line risks—and others.

At the time, the number of ships hijacked numbered less than 1%, and violence was not present, according to an early profile at the time in Wired magazine:

https://www.wired.com/wp-content/uploads/archive/images/multimedia/magazine/1707/Wired1707_Cutthroat_Capitalism.pdf

Owning companies were making a conscious risk choice, to transport cargo (and cruise ships!) though a pirate-infested shipping channel. To take an alternate route, transport time and costs were increased significantly.

Costs, attacks and increased ransom demands rose rapidly. It quickly required focused attention to manage the spiraling risk.

Wikipedia reports today that the Somalia piracy rates declined by 90% between its early start and 2012, due to measures implemented: Armed security on the boats, onshore security forces, more naval presence in the area, and better management practices on the vessels. Piracy attempts still occur, and greater risk has shifted to Africa's western coast.

This is a real example of risk where business owners made a choice to accept risks over the tradeoff in costs and transport time. The yacht owners were an exception—and probably ventured into the waters in error.

Risk Decisions

Risk is a decision to be made by choice not default. Our role is to help management make informed decisions.

No, one size does not fit all. It depends on what your business needs, and how you define risk. A hospital that is handling online healthcare records carries greater risk than bicycle shop.

It depends on what data you need to protect and what your core business drivers are. Some people will work to sell you a one-size-fits-all solution. As the responsible manager or hired expert, you must know enough about the business to ask the right questions for the right support.

You cannot outsource accountability or liability—you own it—you and the business.

Assessing Risk

Risk assessment: The overall process of risk analysis and risk evaluation. You are estimating the likelihood of adverse effects that could result from exposure to certain risks in the environment.

It is a subset of risk management.

Look at what you have in your environment or universe that you are responsible for.

Scope your assessment:

If you are assessing a complete environment, it is important to comprehend the purpose and expected outcomes, to budget your time, and manage the workload.

If this is your first risk assessment, just get started, and keep it simple. Many resources exist, but do not look for a complicated or difficult template. Your process will evolve. Just get started.

Scope it to a manageable size, and base it initially on an area of high risk. Overly ambitious risk assessments can create a significant resource commitment—for you and the business groups where you execute them.

Scoping examples:

- Cause of recent breaches or malware incursions
- PII (Personally Identifiable Information) data mapping across your applications
- Disaster preparedness for critical systems
- Identity management
- Third party access, or a specific cloud or third party provider
- Data center access
- Breach response processes
- HIPAA (Health Insurance Portability and Accountability Act of 1996)

Outsourcing is an option, if you have the necessary resources and financial support for a third party risk assessment. Managing a third party also requires you to document requirements, contract for the work, and then manage and oversee the process including deliverables and payment. If you would like a document that tells you how to execute an RFP—Request for Proposal process, ask me for a copy. Mine is an IT process but it's a good simple framing if you bid contracts out.

You can find sophisticated models, but you will get more done if you keep it simple. Risk management standards exist in ISO and NIST to help you frame it. You can score and weight using simple measures such as Low-Medium-High with a specific score for risk and impact. You could add weighting for impact.

Quantitative measures are best—if you can create a numerical value or financial value on the risk. Qualitative measures work too—interviewing people, conducting surveys, creating baselines then measuring change, etc. Qualitative measures are easier but quantitative is most desired. Use a mix.

ISACA's CISM* exam study guide, and the risk management module will give you an excellent perspective, and worth the purchase of the book. Local chapters often offer very inexpensive training and you need not be a member or taking the certification exam to attend. Phoenix offers workshops twice a year—and cheap to fly in for the overview if you live in a neighboring state.

**Information Systems Audit and Security Association, Certified Information Security Manager*

Current Trends Impacting Security Risk

Security risk changes as technology and the world evolve. Technology will always present risks, it always grows and changes faster than IT (Information Technology) and the business can keep up. Some industries run at the edge more than others.

The rapid growth of the technology market creates significant risk, often because risks are not considered up front.

Emerging Technology

- Cloud computing
- Internet of Things (IoT)
- Mobile payment systems
- Smart phones in medicine*
- Wearable devices
- Drones
- Driverless vehicles

*Smartphones are revolutionizing medicine
February 18, 2017 by Jean-Louis Santini
https://phys.org/news/2017-02-smartphones-revolutionizing-medicine.html

Smart phones and tablets are being turned into medical devices. They are being used to diagnosis and treat illness and chronic diseases; track your steps, calories and heartbeat; research their use in clinical trials; and give patients the ability to manage their own care. It is non-invasive and can be completed remotely. Clinical trials carry huge costs and time to market. This platform will help reduce costs and time to market. Smart phones increase capability and access in realtime. It manages doctor and researcher work, clinical trials and electronic health data.

Cloud First: Many organizations are adopting a "cloud first" strategy, especially for start-up or mid-sized and small organizations, startups, and those making upgrades to their infrastructure. In the U.S., Governments are often following this strategy to save money on infrastructure costs—and stay current.

The Internet of Everything: Everything is on the Internet. You often do not know where it is and who has access to what. Surveillance video networks have become widespread and fit this category.

Connected Homes: This weekend's news headlines in USA Today mentioned connected homes as a new challenge. A seller will retain virtual access to the home's "smart" devices, such as electronic climate control, or a remotely controlled door locks. Smart homes are a minority but growing segment.

Compliance: Compliance is a big driver and also a big headache. Laws conflict across sectors, and local, domestic and international borders. Compliance is not the same as being secure.

Big Data: Big data creates competitive advantage but also increases risk in the aggregate. The tendency of companies to save everything increases costs and risks.

DevSecOps: The security aspect of DevOps. DevOps literally means development and operations. IT refers to agile development and collaboration between development and IT Operations on software releases to increase speed and quality. It seems an intuitive oxy-moron, but these groups have not naturally colluded in the past. It is an evolution for how software is implemented in production.

Other trends and opportunities

- Biometrics
- Identity and universal authentication framework (replacing passwords and user ids)
- Web application security
- Digital certificates and encryption keys

Security has evolved over time to a risk based function, rather than an IT focus on technology. Technology remains a security priority—and a top boardroom concern. It provides the infrastructure for the business to run, and it brings significant risk.

Security in the Boardroom

Be careful what you wish for. With the onslaught of public breaches playing out in the media, and the real costs, security has become a top risk for senior business leaders. Despite the high profile of security and increased spending on staff and technology, major security breaches continue at a high rate.

NACD—the National Association of Corporate Directors, in their Public Company Governance Survey listed these as leading trends impacting corporate boards in 2017

- Global economic uncertainty
- Increased regulatory burden
- Significant industry changes
- Business model disruptions
- Cybersecurity threat

Download the executive summary at: https://www.nacdonline.org/store/product-detail.cfm?itemnumber=37388

NACD is a good resource if you want to know what executives are thinking about risk. You need not be a member.

Executives want to know "Are we secure?" They do not want a rundown of our vulnerability scan results. While not all of us communicate in such terms, many of us in IT-land are criticized for an inability to communicate security risk effectively to this audience.

What Can You Do Today

Suggestions for things you can take on for one or many buckets, to manage your immediate risk:

- Attacks: Plan ahead for how you will manage and respond to an attack. Incident identification and management are critical.
- PII: Identify PII—Personally Identifiable Information and separate it. Do not collect data you do not need. Track sensitive data that you need to protect.
- Cloud Computing: Look to the Cloud Security Alliance (CSA) for current standards and guidance. Contract management is very important, and having a back out plan. Start with something straightforward that is not mission-critical. Local CSA chapters exist across U.S. and cities across the globe to educate and engage you.
- IoT: Review your IoT units for potential unauthorized or unwanted access. What data is collected if any, and who has access to what and how? Identity management may be needed, and network segmentation and protection if it runs over your IT infrastructure.
- Compliance: Look across the organization to leverage efforts, and avoid redundancy and conflict. Compliance can negatively impact the revenue stream. On the other hand, it enhances revenue positively for consulting companies!

The negative impact of regulation on economic growth is talked about in "CEOs Present Trump with Regulation Hit List", by Ross Keely, 2/24/17, in Chief Executive magazine. CEO's are responding to Trump's plans to reduce regulations. They are concerned about a cumulative negative impact, and creating bureaucracy without value. Security is not included in their *Letter to the White House on Top Regulation Concerns*. Export controls are—which applies to some of our technology industries and government contractors. Do not lump your program into this bucket.

- Policies: Keep your policies simple—including supporting procedures and standards. Create and keep only what you need. Make the documents easy to understand and find. Ask for a copy of our guide on this.
- Worst Practices: Look for your worst instead of best practices, especially in an immature organization, or one that lacks a strong security program or support. This will work well for any business.
- "Free": Seek open source and public domain solutions. Look at what others are already doing instead of starting from scratch. On the other hand, freeware is not always free. Licensing may exclude commercial organizations, and support will be absent to make it work in your environment. It might break when you update systems. Go in with your eyes open.
- Security Strategy: Create a security strategy and roadmap for your organization—if they do not already exist. Include a quick SWOT analysis on your strengths, weaknesses, opportunities and threats—to your security function.
- Metrics: Identify metrics to measure your program. Different metrics will be reported and measured for different audiences, especially for executive reporting.

- Industry Engagement: Immerse yourself in industry organizations outside of your workplace, to see what others are doing, and what solutions might be out there. Get networked beyond your own walls.

Cybersecurity Risk Management Career Trends

Cybersecurity Job Demand

The job market is experiencing huge demand for security professionals, with demand outpacing supply particularly for experienced mid-level technical experts. Security analysts and engineers seem to be the most sought-after positions. Many risk management roles exist as dedicated hires or contract consultants.

Contract roles are available readily in most markets, to satisfy peak needs, targeted skills, or support outsourced services. Recruiting firms and Security Service Providers (SSPs) readily fill this busy niche.

Cloud computing is an emerging field where high growth is occurring. I regard this as a shifting of the workforce not a net new employment category. However it will require new skills—it is not just IT as the same old business model. Cloud has a major contracts and contracts management element that very much impacts how your services work. Certifications and training exist to support you if you choose this field.

Job openings in cloud include sales, which is similar to any technology sales role. Infrastructure engineers support cloud companies offering direct client support. Full data centers also hire staff to support customer cloud computing contracts in IT operations, security, and contract management including solution architects. Cloud security architects, engineers and consulting roles dominate Indeed's current cloud listings. Cloud, big data, analytics and software development are often tied together. Google, Amazon, Microsoft, Verizon, Accenture, and a gaggle of other employers seek these hires.

Security-as-a-service is gaining in popularity. Along with growth in third party providers, comes a growing focus on cyber insurance policies and how to shift risks on costs, fault, solvency and litigation.

IoT is widespread and growing—evolving quickly to the "Internet of Everything" (IoE). In IoT, you will be supporting an implementation based on a particular organization sector, size and type where you choose employment. Wearables, drones, and driverless cars—those all fit to a market segment for companies making the product. Your security role will fit to the business model at the company where you work.

Mobile is becoming the dominant platform. Mobile and IoT developers are in demand, and nothing would please our field more, than a developer focused on security!

Security Operations need people who monitor security, respond to security events, and identify and escalate incidents based on risk and pre-defined criteria.

Merchant ecosystems are part of supply line risk, and the need for reasonable oversight of third parties.

You can enter with education and certifications, but the market shortage is in experienced skilled professionals. There are many opportunities to transition and some fields work better than others.

Pragmatists Guide to Working in Security

For those of us in the field, we are always pushing the walls. Sometimes people batter you, want you to say yes on everything, that everything is fine, and there is no risk. It is a little like your dentist—nobody likes going there and we do not want bad news.

Security is often unappreciated. Numerous IT managers will give a talk at conferences on how to manage the CISO Office and fix all the security people working for them, who are smart, but do not know enough about the business. You may fall under attack for being non-business friendly and someone staffed who only wants to make their job harder. That can be the perception from those around you, procurement folks, and business groups, and often your own CIO (Corporate Information Officer) or a new incoming VP (Vice President) level CISO. It may be true—but it is also a party line. The new generation runs into the old, and the new have been told that security is like this, and they must fix us.

You may be hired solely to meet a compliance regulatory requirement. You may take the fall for a breach—and maybe you were actually at fault. Like the Energizer Bunny, you keep on going. You come to work, and do the best you can. It sometimes can feel like being a tax agent. Do not tell the business no, tell them how to do it with less or no risk, or discuss the risk with them.

Security has been ignored and pushed far down in the organization for a long time, and chosen first for IT cutbacks (where we often live). Now the spotlight is on us, and many of us have grown used to working in the dark. Some of us will thrive and some of us are more comfortable in the dark and not meant for the big city and bright lights.

How to Get a Job in Risk Management

Audit Departments

Start by talking to the IT audit director for your company. You can also attend an event at your local chapter for ISACA –the Information Systems Audit and Control association, and meet practitioners for companies in your market. Listen and learn.

Arrive early and stay late, and sit up front so you can meet the speakers and chapter leaders. Chapter leaders can help direct you—they will know the field and the local market.

Internal and external audit roles allow you to work directly in risk management, especially in IT and security. Roles: Audit managers and directors, and auditors that specialize in IT, security and vendor assessments and other audits. All audits are risk-based and focused on compliance to policy, operations procedures and regulatory requirements. Go to ISACA online for job openings and information.

Consulting Companies

Consulting organizations provide outsourcing support, and offer great starting places for those with a basic background. They present hiring opportunities, in organizations such as KPMG, Ernst & Young, Booz Allen Hamilton and many others.

These are focused on supporting compliance and regulatory requirements, internal and external auditing, or security program planning and deployment. The breadth and scope of their work depends on the company and size. Consulting businesses represent a significant piece of the market including contract 1099 hiring. Entry level positions are available, although most require a strong academic background.

GRC—Governance, Risk and Compliance: "GRC" is a common designation in larger companies. It typically covers security strategy, the security policy process, education/training/awareness, audits, risk assessments, compliance activities, metrics, the use of specific tools or databases to support these, audit interface, and management of the security web site presence. GRC is usually housed within a security organization.

"Compliance Analyst" and "Risk Analyst" are common job openings at financial companies outside of audit functions.

Enterprise Risk

Enterprise Risk Managers focus on total risk, not security or IT specifically. Supply chain risk fits well to security because many suppliers are IT and security vendors. Third party risk assessments are common in highly regulated industries, such as banking.

Disaster Recovery: Fits into risk management, along with Business Continuity Planning. Visit the ACP—Association of Contingency Planners' web site, and a local chapter.

Privacy: While the Privacy Officer will usually require a law degree, many other openings are available, such as a privacy analyst. Privacy is a growth area. Visit the IAPP—International Association of Privacy Professionals—web site. They also list job openings under Career Central and you can subscribe to this job listings (free) newsletter without being a member.

Other Opportunities: eDiscovery and forensics, fraud, physical security and investigations (the ASIS organization), and the federal sector of the U.S. government.

Many positions and skills will transition if you can demonstrate a bridge for your own skills and experience.

Certification and Education

Certification: ISACA offers a CRISC certification that can make you more marketable. Certifications represent investments in your career, but proceed with caution. They carry an upfront exam cost, and with the CRISC, annual maintenance requirements for continuing education and a maintenance fee. A certification resource guide is listed in the back.

College Entry: If you plan to enter the field from college, work as in Intern in the field. Policy technical writing can present an entry point here, or more specialized roles based on your background.

Get Involved

Join organizations like ISSA: The Information Systems Security Association or other local security or specialized groups will give you support and education. Be active. Groups exist for most specialties. They help you build a network, educate you, and connect you easily to expertise and a ready job market.

Help yourself and others by volunteering in a board support role.

Technology as a Career Choice

A technology career is a good choice to make—especially for women. It gives us the freedom to live our lives by choice. It pays a good income that lets us make our own choices for having the kind of life we want. It allows us to make substantial contributions to business and the world. Ask me for the Women in Technology Success Manifesto that we created recently.

Women in the Workplace

Some companies have created women's internal network groups to support their female workforce—which are also open to men. Others like Intel have created a more focused approach to ramp hiring and retention of women in technology.

General Electric is working on gender parity in their entry program for technology hiring. Posted in Fortune magazine in the 9-Feb-2017 edition by Lucinda Shen: "GE Wants 5000 More Women to Launch and Build Their Tech Careers There."

In my security and IT tenure over the years, I feel my opportunities and pay have not differed from my male colleagues. It is nice however, having a women's network for moral support and friendship among peers.

If you start to feel alone or different, create a network for yourself, and get involved. Focus on the job and not being different. That is the advice given by the Katherine Johnson mathematician character in "Hidden Figures" story, as noted in "Here's How I Stopped Worrying About Being the Only Woman in the Room", by Mary Godwin, 7-Feb-2017, Fortune magazine.

Women Networking Events, Groups and Participation

Much of my networking time and support have been focused toward female colleagues –women in technology. Support is not exclusive to women, and much translates across gender. You can try different groups for what works best for you, and move on when one is not a fit. You learn from them all. Here are mine, as related to women. I've been involved in many others, and ISSA has been my most consistent investment.

Women in Technology Mastermind

Creating a small group environment was more effective than one to one relationship building, for economies of scale. For professional women in technology, it worked well in a boardroom group mastermind. By invitation only, the mastermind was targeted for those who desired to develop a network and greater results. It focused on positive individuals who took ownership of their careers and lives.

The mastermind forum offered an accelerated way to drive it forward at one's own speed—with help. This kind of network adds up to more than the sum of the parts. Knowing all the players myself—how they got invited!—helped to connect women outside the forum, whether they participated in one or all, or met physically or not.

However, you get as much as you give. Some people participate only to take what they want or need at the time, and then they disappear. The benefit is to build and maintain trust-based relationships that work two ways and play out again and again over time. You actively participate and you actively give of yourself.

ATW Phoenix Board and Conference

When the Alliance of Technology and Women was an active association, I joined the local board, under two different leaders. For 5th and 6th graders in local schools, I led the Great Minds program. We arranged for a monthly female speaker series focused on our field. It was designed for girls, to grow their participation in our field, but school administrators required that it be presented to the entirety of the classrooms. We were still able to influence the girls—and the boys too. This program did not scale—every added school required an equal increment of coordination and time, plus onsite visits. You were performing this volunteer role while working full-time, towards a schedule that could not be maintained. Sponsorship was the second role I took on with a co-chair, who also became a long-time associate and friend. A missed opportunity at the time, was stepping up to lead this local chapter before it died off.

These groups die when you lose good leaders, or fail to attract and manage a strong leadership team. Most are reliant on volunteer leaders. Just like in the "real" world, we are not all created equal, in contribution levels, skill, passion and availability. Not in ATW, but in many association boards, people may be seeking the power of the position, rather than passion or work. Someone has to do the work—it does not happen by itself. Volunteers are typically professionals in their own right; they are not subordinate or paid staff members. The team must be managed and led to their strengths, but also for accountability. People contribute at different levels. Some show up fully, and others never show up at all.

You learn new skills, and most importantly, you build strong relationships from your board colleagues, and the active members. You gain much more when you fully engage as part of a leadership team. Seek out opportunities where you can contribute and enjoy the experience.

Women in Technology Conference

Late last year, I organized and hosted a local conference for women in technology, under the ATW brand. Although it received raves, it required a large amount of time and effort to pull the event off—financially and to get the attendees there. Everything is a tradeoff on these events—time, cost and participation.

Women's events are often perceived as fluff. Some managers do not support them, nor do some women. Competition is fierce for our time and attention, away from the office or home. I do not usually announce that I am attending a women's event—unless it is to another woman. These often are not well received by male colleagues, who feel they are being excluded (when they cannot do the same). However, we do welcome men. Content is more specific to female concerns but anyone would benefit.

In round table focus groups, we created a Women in Technology (WIT) Success Manifesto. Ask me for a copy if you want to share insights from this large group of female professionals—and men.

ATW Conference in Phoenix, Arizona

EWomenNetworking.com Phoenix Affiliate

In Phoenix, upon leaving Intel to start a small business, eWomenNetwork approached me to start a local affiliate for small women business owners. Yes, was my response—and eWN was launched in Phoenix with me as Managing Director. I believe that my NSA membership—the National Speakers Association—was used to identify me as a target—and others.

After a year, and looking at time invested vs. financial returns, I made a decision to move on. Revenues came from membership fees and selling onsite exhibit tables, which netted little to the person bringing them in. Catering was breakeven or subsidized, for event attendance. The affiliate netted me a consistent loss, while earning revenue for the company. It was not a matching target audience for me.

eWN recruits you to lead Affiliates through a sales person who gets a substantial commission for you joining. It feels like an interview for a "job"—but it is a sales job, and you are part of the money network—theirs not yours. You pay an amount up front to become a "Managing Director", and keep paying monthly for the privilege of membership, just like the members. It offers a win-win for everyone but you.

You are responsible for arrangements and costs of the meeting venue, catering food and beverage, signage, collateral for the event, and all the work to execute. The web site was hosted by the company, and you were reimbursed for event registration fees paid by attendees, minus transaction fees. Well-crafted meeting scripts were provided by the organization, and registration took place on their site (to control the contacts and the money that you brought in).

It acted like a giant Network Level Marketing company, selling local affiliates and making a profit on all your work, while you took a loss. Some managing directors did make money, I assume. The concept was good, and the management leaders were marketing experts. They created a success recipe for themselves that was refined over time.

EWomenNetworking.com is a for-profit privately owned family business. Some people love it, and if you are good at sales and enjoy building relationships, you could make money by working it full time. Sales are the heartbeat of any organization to bring products and services to market. If this fits your business model, it might also work for you, even as a leads generator.

Getting Paid for Association Work

In doing and enjoying so much volunteer work for non-profit association groups, it became a goal, to be paid for some of my time. Most compelling? Become the (paid) Executive Director of ISSA. ISACA also opened three staff positions when they launched the CSX certification program and interviewed me extensively. Neither worked out for different reasons, and moved me toward other opportunities instead. Engagement in the field continues to offer rewards beyond money for the relationships and skills grown.

Joining Support Groups for Women

Over networking get-togethers outside of office hours, our local women also participate informally in women in security group. Within ISSA, the Women in Security SIG—Special Interest Group (WIS SIG) is active. These support other women in our workspace, not for man bashing, or even to talk about men. We talk and collaborate as female colleagues and friends. Women are fewer in our individual work environments. It is helpful to meet and build relationships with others like ourselves.

Growing the Pool of Women

Technical conferences in our field typically end up with a majority of speakers that are male—sometimes all of them. When women recruit the speakers, there will be at least some number of females on the agenda. If you ask the men to look for more women, you will be reminded to find the most qualified person, not look explicitly for women. Audience make-up will reflect this disparity also—depending on the number of females involved in the organizer's leadership team. This is true also for association membership and leadership.

Part of my goal is to consciously think of and seek out women in my network for opportunities.

Last year when attending the Phoenix IoTDevFest—for the Internet of Things—only three women were in the audience when it started, and only five attended over the day. One was a female presenter—a Hardware Electrical Engineer. This crowd develops for the Internet of Things. It helps to educate oneself beyond security in this growing segment of connected devices on the Internet. Opportunity knocks ladies.

IoT DevFest Mesa, Arizona

Whether you are responsible for speakers, or leading a board, or seeking writers, work to increase female numbers. More women lead to more women. We reach out to others, who then reach out to their friends and colleagues. Then women start to show up more in the speakers, audience, leadership teams, and in hiring.

Conclusion

Step up. Share your expertise by writing, speaking and leading. If you recruit speakers, actively seek more women. Reach out to network, connect and build relationships. Get out and be seen and heard, and help others create a voice.

Grow your credibility and the presence of yourself and other women. Take the opportunity to participate, and share your expertise through speaking, writing, leading and reaching out to others.

Bonus: A Male Point of View On Women in the Workplace: Interview with Kim Jones

Preface: On a women's panel that was facilitated for an ISSA security conference, I noticed that women do not articulate why they have separate forums, why a women's group will help them, and why women might need to do something different for themselves. I wanted to get a male perspective on why this is so.

Kim is someone I know and respect, a longtime security executive, who supports women and the unique groups for women to build their own support network.

Kim L. Jones, CISM, CISSP, M.Sc.
Director, Cybersecurity Education Consortium
Arizona State University

<u>What do you think women's unique problems might be, based on your extensive management experience?</u>

Kim attempts to provide mentorship within his female staff members. He also mentors colleagues in the field, both male and female. He can teach what it takes to be a good security person, a reasonable successful executive, but cannot teach you how to be a woman in our field.

<u>Why do you think that women need it to be different?</u>

Kim states that whether you like it not or it's fair or not, there is a distinction between genders in our field, including pay.

<u>Do you believe it is really a problem?</u>

Hell Yes! Women to a certain extent, are still looked at as second class citizens. The concept of true neutrality across the board does not exist as much as we strive for it.

Even in the most enlightened organizations, the capability for folks to succeed is not a truly equal playing field. In technology, women are the ones most often disenfranchised, in similar fashion as they have in the military for many decades, up until recently.

He cannot explain why; it is a geek version of machismo. Technology or the matter of hard sciences of men vs. women have led to this being a predominantly male career path. It seems trite and stereotypical, that a woman who portrays herself demurely, is weak, and who portrays herself strongly is a bitch. (The stereotype). A person who carries themselves like Kim might be also be labeled negatively.

He can understand these academically, but has no idea how to walk that wire that a senior woman technologist has to do. It is ironic, that women typically succeed more than men, due to high touch.

Statistics proved that out, that more women in technology who were gravitating to security, were succeeding statistically more than men.

At the time, it was believed to be because women had a better grasp at soft skills—especially at an executive level. Women were able to strike that balance. (These skills are often perceived as a weakness in our security leaders.)

What do these groups—for women in technology, women in security, or a group mastermind for women—what do they bring for women? Not all people support that uniqueness. How does it help women overcome barriers?

First, it helps to focus on uniqueness that exists, to drive that uniqueness out (between men and women in the workplace). It helps equalize the playing field for women.

Second, it gives a level of commiseration and validation of the concerns.

Women in technology are a substantive minority. She has only her own experience to deal with. The ability to hear someone else articulate your problem in a way that resonates with you, gives you that sense you are not alone. It gives you that level of mentorship to get through the challenge, which is huge. Kim hopes a day and time will come, with this level of separation is not needed.

Kim advises women joining a special interest group to make the most of it for themselves. Joining without participating is fruitless. We create the learning and networking opportunities. Join, show up, and participate.

Risk Management Conclusions

Security is all about risk. Information security decisions should be based on organizational risk. Everything is a tradeoff on opportunity cost, risk, operational impact, and resources required to manage risk. Risk is a board owned activity—or at the most senior level of the business. Our role is to help the organization make conscious informed decisions.

New technology is nothing new. Risk tradeoffs present business opportunity choices. Criminals follow the money—and today, the money path is electronic—on systems and databases. You can easily prevent most breaches. Focus on high risk and high return for your business.

Step out of your comfort zone. Do not wait to be told you can do something. Take charge yourself and open your eyes for opportunities to create value where you see a vacancy. There is always more to be done, that no one is doing.

In career planning, always be building new skills in the job marketplace. Due to the risk, career and job opportunities exist in cybersecurity across the board. The door is wide open like no time before.

Thank You for Being Part of Our Story

Thank you for reading our stories. Thank you to the women who took time to write and share theirs.

Each of us impacts the lives of others. Make time to tell someone about a positive impact they made on yours. Engage more women in your hiring pipeline, and actively recruit women for your speaking opportunities and for leadership roles.

Ask me for a copy of the Women in Technology Success Manifesto.

Every day is a new day. Help create technology solutions, and help create opportunities for those people and the world around you.

Resources

"Jobs of the Future and the Two Skills You Need to Get Them" https://www.weforum.org/agenda/2016/09/jobs-of-future-and-skills-you-need/The key is combining math and interpersonal skills.This gives an excellent overview of what we need for the future workforce.

General security job roles (many others exist, but these are core):

- Information Security Analyst
- Information Security Architect (typically an IT architect or developer)
- Information Security Engineer
- Information Security Manager
- Risk Analyst
- Compliance Analyst

Government job openings may replace the word "Information" or Security" with the word "Assurance" in the job title.

Some of the main association groups that support security (and privacy)

- ASIS (including a Women in Security Council)
- CSA (Cloud Security Alliance)
- EWF Executive Women's Forum (owned by Alta Associates, a recruiting company)
- IAPP—International Association of Privacy Professionals
- ISACA—Information Systems Audit and Control Association (includes a women's group)
- ISC2—International Information System Security Certification Consortium
- ISSA—Information Systems Security Association (Including a Women in Security Special Interest Group WIS SIG)
- SANS Institute: Well respected but expensive information security training organization (for profit, not an association). They also have a women's initiative, and certifications.

A sample of common security certifications (hundreds exist!)

- CCSK—Certificate of Cloud Security Knowledge
- CEH—Certified Ethical Hacker
- CISA—Certified Information Systems Auditor
- CISM—Certified Information Security Manager
- CISSP—Certified Information Systems Security Professional

Vendors also offer non-neutral certifications that can open doors and do not require experience. Network firewall certifications can be extremely valuable (Cisco, Palo Alto, CheckPoint), SEIM products, and risk management database software.

Why Certifications?

Certifications are often requested in job requisitions, to demonstrate a base level of competence in a given field. CISSP is one of the most dominant, but many exist, and they vary in value. Certifications can help you get hired, and they can also help you build and maintain a level of knowledge beyond your current state.

When you decide to get a certification, consider why you want it, what its value is to you and the job market, and what the maintenance requirements are. Every certification represents a real investment, at a minimum, in preparation time and materials, plus a financial commitment—up front, and often an annual renewal fee.

Many certifications exist in the security field, and they vary in value, criteria and cost. You can find one that fits your own goals (or not). In security today, you can be hired without a certification, but often you will find one or more listed as a hiring requirement.

For my own certifications, CISSP* and CISM* require a base level of experience in addition to passing an initial 6-hour and 4-hour respective exam. If you fail to maintain required annual maintenance in education hours or the annual fee, you must pay for, retake and pass the exam again. Otherwise, the certification expires. For a period of time, I also carried the CIPP*/IT credential.

CIPP and the IT exams were separate, 2-hours each, and required no experience, just to pay for and pass the exams with a passing score. At the time, IT represented one of three specialties on top of the base privacy credential: Government, Europe or Information Technology (IT). IAPP changed the certification naming convention and combined these two into the CIPT later. Finding less day-to-day value for my own career in this certification, I allowed it to lapse.

*CISSP, ISC2's Certified Information System Security Professional, is one of the most recognized and requested certifications for cybersecurity professionals. (Ask me if you want a copy of my paper on this certification and best practices to prepare for and pass it.)

*CISM, ISACA's Certified Information Security Manager has grown in demand over the years and with cybersecurity prominence in the job market. (If you want tips on this exam, ask for my written report. I facilitate the workshop for our local market.)

*CIPP: IAPP's Certified Information Privacy Professional and related certifications have grown along with the emergence of privacy as a profession.

*CIPT: IAPP's Certified Information Privacy Technologist.

My personal thoughts on certification:

- They can improve your hiring opportunities and make you a stronger candidate.
- If you are the hiring manager, you know there is some level of competence (it varies!).
- While not all certifications create tidal waves of profits, many do. The certification market is an economic engine that generates huge revenues for the organizations selling them. "Certification" is a money-making product for sale. If it is a for profit company selling certification, it is offered to help sell or support their product. Examples: Firewalls, risk management databases, security event management systems, or HIPAA (Health Insurance Portability and Accountability Act) training and services.
- You do not need 15 certifications next to your name—pick and choose wisely.
- Some people only obtain certifications if their employer pays for them. Your career belongs to you—it is up to you to invest in yourself. This will become clear if you find yourself suddenly unemployed, or if you work as a self-employed consultant—or if you are the paying organization.

This is another opportunity to give back while you also learn—offer to lead a study group on a certification that interests you.

Why a Career in Cybersecurity?

Security is an evolving field with constant opportunity and change. You can earn a good living, doing important work, in a red hot field that is projected to keep growing for years. Demand currently exceeds supply. As you see from this tiny sample of women in our field, you have the opportunity to choose from a wide variety of responsibilities and roles across the government or private sector, in large or small organizations, and local, national or across the globe.

Chapter 4
The Era of Homo Digitus

Evelyn de Souza

We have emerged as a new species. We have become digital beings whose lives are closely anchored to digital objects: Deep learning, artificial intelligence, and behavioral analytics are no longer buzz words; they are part of our new reality. From a security and privacy perspective, there are new and different considerations. That's why it's fitting to focus on the concept of Homo Digitus, which I first learned about in "*The creative destruction of medicine: How the digital revolution will create better health care,*" by Eric Topol, and more recently highlighted as a Gigaom conference theme.

This chapter expands on a concept I wrote about in CloudTweaks last year In Topol's vision there is a new human species, *Homo Digitus* that benefits from the data deluge brought about by the convergence of the digital and physical world. They track sleep quality with brain-wave headbands, monitor vital signs with wrist transceivers and use cell phones for self-diagnosis amongst other things, realizing the opportunity for a much more evolved life.

In this chapter I will focus on the following from a security and privacy perspective:

- Technology Transformation for Homo Digitus
- Safeguarding Data, the Lifeblood of Homo Digitus in the Enterprise
- Gender Evolution in the Era of Homo Digitus

E. de Souza (✉)

D.A. Christofferson (ed.), *Women in Security*, Women in Engineering and Science, https://doi.org/10.1007/978-3-319-57795-1_4

Technology Transformation for Homo Digitus

Each evolution of the human species delivers on the promise of a smarter human far more capable than the species that preceded it. In earlier species humankind looked for new ways to improve hunting and gathering of food and in some parts of the world, such was their mastery that they even created an oversupply of certain foods.

Homo Digitus sees data as the new food and you'd be challenged to find any facet of life that has not been revolutionized by data deluge—there are a great many varying estimates, on how much data is created every year but everywhere you look volumes of data are soaring.

Mostly, it's been for the better good of humankind. Hands-free control of technology improves the lives of people with physical disabilities, makes healthcare and other professional services more efficient through accurate voice dictation, enhances automobile safety, and makes everyday tasks more convenient.

With new data points, we've also experienced great transformation in the quality of services and business process too. We've come to have a great reliance on the benefits that data deluge brings. If you consider how often an individual might use map services to estimate a commute, or an app to compare shopping prices, and then for business the productivity and economic gains from using data points to estimate customer preferences, improve customer service and speed business delivery times.

Will Your Internet of Things Device Testify?

However, a strong note of caution and a real-life scenario: Your wearable device is subpoenaed to testify against you. You were driving when you were over the legal alcohol limit and data from a smart breathalyzer device is used against you. Some might argue that such a use case could potentially safeguard society. However, it poses a bigger concern about how data from the broader spectrum of connected devices or apps could be used against you.

Doesn't it seem reminiscent of George Orwell's dystopian universe, "*Nineteen Eighty-Four*" where children were indoctrinated to inform on suspicious activity, only now it's a connected device? But, this time it's you who chose to use the device or network of devices, or apps that could start working in concert against you.

Some might see this as the gradual erosion of privacy, but in the era of Homo Digitus, privacy will be redefined. There won't be a universal single definition; each person will have their own privacy threshold about the level of data they feel comfortable sharing with third parties and about the implications of that for their futures.

Controlling Data Deluge to Maximize on Homo Digitus

One of the downsides of the era of Homo Digitus, is that data has been exploited and misused in ways that leaves humankind fragile and that puts businesses at great risk. Data breaches, privacy infringements, identity and data theft abound.

When new technologies are released that introduce microphones or other recording devices for the first time, or seek to use data in new ways, companies need to provide additional transparency and greater levels of control and choice, and in a way that consumers can readily digest. In particular, we need to look out for older people who technology might be anathema to and children (see further on) who may run the risk of being exposed or harmed in ways unintended.

Until these standards are in place, I encourage users to be vigilant and to press manufacturers for clear answers on the following at minimum:

- **Will your data be shared with third parties?** This is particularly important for any device that collects sensitive data about you. It may be challenging given the volume and legalese of privacy policies but well worth the time investment given it's your private data.
- **Understand how your information is transmitted.** And, once in storage, who has access to the information? Is the information stored on a third party's cloud? When you stop using the device, what happens to your data?
- **Take time to understand your device's privacy settings**. Have you configured the device's settings maximum privacy? Are you only sharing what you are comfortable sharing publicly?

Protecting Our Next Generation

Unlike in the physical world, danger is far more difficult to discern. In the physical playground, an object hurtling in the direction of a child might be the cue for a child to move away. Whereas in the digital world events that may result in financial harm of personal exposure may occur stealthily over time or suddenly without warning. Parents tend to be more watchful of their children in a physical playground and teachers watch over children during recess and lunchtime. While many parents are very diligent and keep a steady watch over their children when they are on online, it can be challenging to monitor everything a child does online given its pervasiveness.

When children use computers, tablets or mobile devices, they may think that their information is stored somewhere on the device. And, as savvy as they may be with navigating their way around a device, they may not always realize that their data has been silently and without any cue transmitted to cloud storage.

The good news is that in many parts of the world, privacy laws are catching up with many cloud-based applications. Regulations are increasingly requiring parental consent for collecting certain types of sensitive data and prohibiting the use of children's data for marketing purposes. Regulators are also cracking down on deceptive practices, especially as they pertain to children and those who may be vulnerable.

However, until we have consistent regulations and standards, parents, adults and teachers need to exercise even greater caution and oversight as compared to the outdoor playground. It's the reason why online safety classes needs to be a standard part of all children's education today. And, it's the reason why I am focused on accelerating digital safety initiatives, not just for business usage, but for everyday living also and especially for a safer playground for our children, especially my niece and nephews.

Safeguarding Data, the Lifeblood of Homo Digitus in the Enterprise

What's needed is a way that allows Homo Digitus to benefit from the positive effects of data deluge but with the safety net that data flows are being directed to only those authorized to have access.

In the workplace, digitization has changed how we work—it goes beyond the devices we use and where and when we work, and more to the tools and data and our interactions with an expanding network of people and data. Yet, despite the fear around security breaches, there are few security approaches that truly focus on securing at the data layer with a contextual focus on people and the expanding number of applications in use today.

In an enterprise being able to proactively determine data flows and then implement additional safeguards based on a comprehensive set attributes ranging from geolocation, network, and device down to multiple facets of identity will be critical, given the sophistication of data exploitation.

Information security tools have traditionally been associated with impeding progress. Newer solutions need to be easy to implement and policy attributes stated in a way which are business-consumable, so that business leaders standing up new services can easily understand security and privacy implications versus having to navigate security speak.

However, as great a risk might be posed to businesses from well-intentioned individuals, who might accidentally misuse or overshare data. It's only natural considering that for many individuals digitization occurred midway during their lifespan and dealing with the data deluge is not yet a completely natural phenomenon. Also, our lives are increasingly fast paced, workplaces are more pressurized and the convergence between home and work narrower, that workers are prone to accidentally misusing data.

For example, it's very easy to accidentally drag and drop a file into an email and send as an attachment. Toolsets that can alert users to a potential data misuse can greatly alleviate this situation while allowing users to perform their work duties in an efficient way.

Cloud and Data Multiplication

Cloud computing has become the term du jour in the industry and its evolution has enabled highly scalable and elastic services via the Internet and cloud has become the back end for so many aspects of our daily and work lives. Organizations leveraging cloud services to store this data may need to take a closer look at the lifespan of the data they collect and how it is expired and destroyed.

Today's organizations need to understand that cloud as a model causes data to multiply further. The dynamic nature of resource allocation and maximizing availability in a hybrid or public cloud means resources are replicated and backed up across multiple data centers. While there isn't a universal blueprint for protecting sensitive or mission-critical data, the following is a good starting point.

- **Tag all sources of mission-critical data:** It starts with strong preventative measures: If data is classified digitally to a scheme that is intuitive to your cloud provider and your organization it will be easier to track through its lifecycle and then expire and destroy.
- **Take time to assign entitlements and access rights:** Ensure that access rights or entitlements for sensitive or mission-critical data are limited to only those who have a legitimate need for access.
- **Apply encryption based on context:** When data is encrypted, it is only readable to those with access to the encryption keys. It is the most certain way to limit unauthorized access to data in the cloud. By encrypting organizations can be better assured of the confidentiality of their data and potentially be less concerned with their cloud providers' data destruction methods.
- **Perform data wipes:** Many government and industry standards require data storage wipes to ensure that hardware is safe for reuse. There are different types of software and hardware that even allow for remote erasure. The benefit is to enable a provider or enterprise to repurpose the media for reuse.
- **Physically destroy data and media:** In the cases of highly classified information organizations can use strong magnets to destroy data or even shred physical media. This ensures that the data on the destroyed media can never be recovered. Physical destruction methods are the last resort and only feasible in a private cloud environment.

Gender Evolution in the Era of Homo Digitus

Though the industry has seen the rise of this and other great technological revolutions, the gender evolution is moving much more slowly. Bridging the gender gap matters.

How can solutions for our future be architected primarily based on the input of the XY chromosome and without input from the XX chromosome? It's the Yin without the Yang. While there are some impressive female role models today in the digital industry, having women in more balanced numbers across engineering and product functions will lead to more innovative and functional solutions. Research suggests that stronger links between left and right hemispheres in women makes them better at intuitive thinking and demonstrating flexible attitudes.

As you embrace digital living, be sure to think about ways you can help bridge the gender divide:

- Encourage girls and women who are interested in technology to seek the support of organizations who mentor and nurture girls and women in technology
- Attend girls and women's' technology events to demonstrate your support for stronger gender balance
- Find ways to attract women as you look to hire for engineering and product development roles
- A key finding from a recent study "Addressing Gender Gaps in Teens Cybersecurity and Self Efficacy" was that teen girls were likely to develop confidence and interest in cybersecurity through informal approaches. It's a great opportunity for cybersecurity practitioners to become role models and mentors to a younger generation. I noted earlier that many cybersecurity approaches lag as much as 10 years behind the business landscape. Overhauling industry approaches is difficult when approaches and toolsets have been in use for decades. That's where reverse mentoring can play a role. Partnering with young people is not just about them learning from us; it's about what we can learn from them.

Homo Digitus is still in the process of being shaped. Now is the time to ensure that data deluge remains the positive enabler. With each evolution of living our lives have become much more strategic and filled with interesting possibilities compared to previous generations who may not have had as much choice as we have. It's time to drop the fear-based messaging. That would help us focus on technology as an enabler and to address issues that are gray or unclear versus fear-mongering which often serves to hold people back.

Chapter 5
This Girl Wants to Go To School Here!

Ilene Klein

You need to know a lot of things to be a security leader. Growing up, and even into my 20s, I never wanted or suspected this would be my career. Looking back, I'm truly surprised at what I've done, how lucky I've been, and the lessons I've learned along the way.

Getting There

In the summer of 1978, right after graduating high school, I took my first college computer classes: *Computers and Humans* and *Fortran Programming*. I hated them! As soon as I wore down my parents, I switched majors to theater performance (!).

Realizing I'd never be employed as an actress (I'm not that talented), I dropped out of school and got a customer service job for a company that sold biomedical research instrumentation. These were refrigerator-sized tools for things like nuclear medicine. I handled service parts. A customer would call and tell me they needed a new potentiometer or transformer or another component I'd never heard of. I'd look it up in a parts catalog, get the product number, and tell the customer the price and whether it was in stock.

But then there were customers who weren't sure what they needed. I ended up bothering the service technicians a lot. One of them (who I happened to be dating) finally said, "You're telling customers to buy transformers and PC boards, but you don't know what they are!" He was going to school to get his electronics engineering technology degree (same amount of math and physics as plain electronics engineering, but more hands on). He literally took me by the hand, sat me down in front of the school's dean, and said, "This girl wants to go to school here."

I. Klein, CISSP, CISM, CIPP/US (✉)

D.A. Christofferson (ed.), *Women in Security*, Women in Engineering and Science, https://doi.org/10.1007/978-3-319-57795-1_5

I started with Electronics 101. Surprise—I liked it! I ended up quitting the job and going to school full time. But I didn't get my electronics degree. With just a couple courses shy of my Bachelor's degree, I ran out of money.

I got a job teaching electronics at a small college. The school's president told me that I didn't need the degree—I needed "demonstrated knowledge," and his school met the required percentage of degreed instructors. After that I worked as a systems engineer installing proprietary software into military bases around the U.S. and western Europe, a UNIX systems administrator, a QA (Quality Assurance) analyst supporting the Department of Energy, and a technical writer contracted to a Fortune 10 consumer products goods company. In a strange way, I worked my way up the "computer stack" from electronic components, through operating systems, to applications and user interfaces.

Then it happened. While working as the CPG tech writer (they needed somebody who could read UNIX shell scripts and explain them in writing), I talked to the head of information security. After he learned of my background, he asked why I was working as a tech writer and hired me.

Although I'd done security stuff as a systems engineer and sysadmin (system administrator), that was my first official job in security. And it was a great introduction to the many aspects of security. I provisioned user accounts, wrote policies and procedures and control frameworks, acted as the backup firewall admin for DR (Disaster Recovery) tests, wrote the risk assessment when the company was considering giving all employees internet access (for the record, my recommendation was not to give access due to the potential for breaches of confidential information and user privacy risks, but management accepted the risk and opened access), and developed the company's first web-based security awareness training program.

This was a busy time for info security.

For example, Microsoft issued security patches as needed—not on their current second-Tuesday-of-the-month schedule. It always seemed like we had a fire drill when a patch came out. What's its criticality in our environment? How soon do we need to apply it? Will it break applications? What do we do?

I put together a process to manage patches and vulnerabilities, presented it to management, and was given a team of SMEs (Subject Matter Experts) for desktops, servers, network, applications, and manufacturing human machine interfaces. Together we eliminated the fire drills and implemented a vulnerability management program.

Malicious software, viruses and worms also exploded during this time period with SQL Slammer, Nimda, and I Love You. We had the same type of fire drill every time we got a malware infection. So I did the same thing—put together a proposed process, got management support and a dedicated team. Together we implemented an incident response program (there were still some fire drills).

Then right after CA 1386—the country's first privacy breach notification law by the State of California—went into effect, one of our vendors had a privacy breach. Because I led virus incident response, I got pulled into the incident and started learning about privacy.

Since working at the CPG, I've moved on to other organizations to help build their security programs (or fix dysfunctional ones).

I'm lucky because I grew into security and privacy as they were maturing. For example, the first security framework I read was the British Standard BS 7799. That evolved into ISO 27001. I learned about VISA's Vendor Information Security Program before it evolved into PCI, and I developed my first control framework right after Sarbanes-Oxley was enacted before our third-party auditor had determined which controls to assess (and I'm willing to bet they "stole" my framework!).

Being There

Even though I've been lucky enough to "grow up" with security, I definitely don't know everything! But I have learned some valuable lessons along the way. I hope these help you in your career journey.

Give Credit Where Due

I firmly believe in giving my team all the credit while, as their leader, I take any blame. Always. I coach and mentor team members who aren't performing (and even fire them if needed), but I never blame them.

On the flip side, there were times when I deflected credit when I should have taken it. For example, I thought up a creative concept and wrote the script for a security awareness program. I then presented it to our VP to approve. When he asked how I came up with the concept, I hemmed and hawed and deflected to the team. I still remember the look on the VP's face. I had just lost all credibility with him. I should have told him the truth, that I started with puns for character names, like Ima Risk, the new employee who doesn't know our security policies, and Harry Arms, who holds the door open for tailgaters. Then I created a series of vignettes structured like and inspired by the old TV show, *Dragnet*.

Learn to Talk and Write "Goodly"

Every day, I use the lessons learned in those long-ago college theater classes. For example, I naturally have the soft, sweet, high-pitched voice of a 6-year-old. A vocal coach taught me how to lower my voice, add inflections, and project so folks at the back of an auditorium can hear me speak. And that's also when I learned to get over my natural shyness to speak in public.

As a people leader, I require my team members to write and give presentations as part of their annual performance goals. I "volunteer" them to write security awareness articles for our intranet site and to give security presentations at take-your-child-to-work day or internal conferences. But I also give them a lot of moral support, examples, and coaching. Writing and speaking well are not only good skills to have (and usually woefully lacking in technical folks), but the experiences give team members exposure to upper management and the rest of the organization. (The term, "goodly" is a bit of a running joke with team members, given my insistence on good grammar.)

Don't Be Afraid to Go Backwards

I was a ***manager*** at a CPG company (oooohhhhh, aaaaahhhhh). Then I took a job as a lowly Senior Security Engineer for a city government (with a significant pay cut). Why? Because it was a chance to help build the city's security program, and I love building programs. The security team for the $3.5 billion city consisted of its first official Chief Info Security Officer and two senior security engineers. My (female) peer built the security operations program, while I built the security and privacy governance programs. And our CISO (Corporate Information Security Officer) kept us in line! We often joked about the "princess to talent" ratio.

After a little over 5 years with the city, I interviewed for and was offered the CISO role for another city. But I turned that down to become the VP Security Governance and Compliance for a leading payment processor. From engineer to CISO or VP—not too bad a jump!

Sometimes You Just Have to Smile and Nod

When I went to school for electronics, I was usually the only woman in the class. Then, when I worked as the systems engineer, I was the only woman on the team (and later found out I was paid substantially less than my male peers, but that's another story).

Travelling to over 270 military bases installing systems, I often encountered local system administrators (usually male) who questioned my competence. My favorite experience was when I visited a base in Germany. I was called in because site management suspected the local system administrator was doing something illegal with the U.S. military systems. This was over a decade before there were computer forensics tools, like EnCase. I had to manually search through files systems looking for indicators of abuse.

The local system administrator was arrogant and condescending. He gave me a smug smile, sat me down at the system console, and logged me in as root. I half expected him to pat me on the head like a child.

All I did was smile and thank him. I didn't get defensive or challenge his authority. I quietly did my job and found the files he thought he'd hidden in the bowels of the operating system.

Get the Credentials

It took me over 20 years to finally finish my Bachelor's degree, online at night while working full time. And for about 3 months, I sat on the floor after dinner with my back against the sofa and used the coffee table as a desk while I studied Shon Harris' *All-in-One CISSP Certification Exam Guide*.

Quick story (with a bit of bragging): ISACA—the Information Systems Audit and Controls Association—holds their certification exams twice a year. I won their Geographic Excellence Award for earning the highest score among everybody in the North American geographical region who took the December 2011 CISM exam (Certified Information Security Manager). I told my boss who spread the word among upper management. Our CIO (Corporate Information Officer) came by to congratulate me, saying he didn't realize only two people took the exam. (Yes, he was joking.)

You need to be smart and knowledgeable to succeed in this field. Degrees and certifications are expensive, time-consuming, usually a pain to get, and don't necessarily mean you know your stuff. (Sometimes they just mean you can pass a test!)

So why go through the bother? They're often the "admission price" for a job, sometimes a set of letters after your name impresses others enough to actually listen to you, and sometimes the achievement gives you the confidence to speak up and with authority.

When I'm hiring security folks, I look for a CISSP, CISM, *CISA, *CIPP, and/or any *SANS GIAC certification. But I also tend to be wary of folks who have too many certs. That gives me the impression the person studies for tests all day instead of working.

(*Certified Information Systems Auditor, Certified Information Privacy Professional, and SANS Global Information Assurance Certification).

Fail Quickly and Keep Going

Along with my successes, I've also had many failures. While I don't believe in doing stupid things, I do believe in failing. If you don't fail, then you haven't taken enough risks. But if you're going to fail, fail quickly, learn from it, and keep going. And don't keep repeating the same mistake.

Yes, I know this is all easier said than done. But maybe you can turn your failures into lessons for others. For example, I've given presentations where I talk about my security awareness programs that failed.

Keep Current

This field is always changing. You ***must*** stay current on emerging threats, actors (good and bad), tools, techniques, and buzzwords. Check out some podcasts and TED talks. Subscribe to (and read!) different security and privacy mailing lists. And read some general business materials too. I occasionally joke that our senior leaders manage by *Businessweek*, so you need to be aware of the hot business buzzwords that often filter down to those of us in the trenches. And be prepared for questions if, heaven forbid, security is a topic in a mainstream news or business article.

Network and Give Back

We can't do this alone. And by "this," I mean be effective and progress in our careers. Join industry professional organizations, like *IAPP, *(ISC)², ISACA, *ISSA, and others. Attend events and chat with your peers. When you develop something that's not proprietary, share it. Share your horror stories and lessons learned.

I've shared policies, processes, and reams of security awareness materials. I've edited peers' resumes, acted as Board Secretary for a local professional group, and I've gotten lots of support, documents, and help in return.

And jobs! That VP job was a direct result of my network connections.

(*International Association of Privacy Professionals, International Information Systems Security Certification Consortium, Information Systems Security Association).

Closing Remarks

Security is a wonderful career field. It's always fascinating, never boring, and sometimes frustrating (so keep your senses of humor and the absurd!). There are some wonderful people working in security and I marvel at their creativity, passion, and brilliance.

Thanks to Debra for including me in this book, and I wish all of you great joy and success.

Chapter 6
Three Decades of Digital Security

Jeani Park

I've spent 30 long years as an exterminator- battling computer bugs, worms, and digital infestations of all types. Sure it's had its moments, but I've primarily played a comic game of whack-a-mole, fevered prayers, and after-action hedging.

Does this mean I preach doom and gloom, certain of cyber war sans relief, all end-of-day dystopian angst? Nah. I'm an optimist, always have been. And when it comes to the world of security, I have reason to be. There are three main reasons why:

- The remarkable advances in computing- particularly vis-a-vis data processing and manipulation of big data, "from everywhere" that "lives everywhere."
- The trend, by organizations of all sizes, to integrate and balance efficiency, automation, and protection.
- The flame of human ingenuity and innovation, with continues to foster unprecedented levels of healthy competition and choice.

Before I discuss how these three factors can and should drive the future of digital security, let's take a little walk down memory lane. I will discuss the advent of popular security products and services, starting just before the turn of the century. Each one came about in direct response to the day's "hottest" security breaches and concerns.

Computer Viruses

Way back in 1990, the first consumer antivirus product was released. But viruses were still fairly rare. It wasn't until 1995, when Windows 95 was released, that the number of Windows desktops consumer and business users exploded. Only a few

J. Park (✉)

D.A. Christofferson (ed.), *Women in Security*, Women in Engineering and Science, https://doi.org/10.1007/978-3-319-57795-1_6

years later, as a part of the Windows 1998 release, Internet Explorer was integrated into the operating system. This, in concert with the release of Outlook 1998, which introduced HTML mail, laid fertile ground for new types of fast spreading computer malware.

Indeed, who could forget the ILOVEYOU worm and the Anna Kournakova virus? The ILOVEYOU worm hit tens of millions of Windows computers in 2000. It spread via email containing a VBScript in an email attachment. Upon opening the attachment, which the user thought was a love letter, the user unknowingly unleashed a hidden VBScript that damaged files on his computer. Additionally, the ILOVEYOU worm sent a copy of itself to all contacts in the user's Windows Address Book.

Computer viruses had gone mainstream, gaining notoriety in popular press. Just a year later, the Anna Kournakova worm hit. This time the enticing email subject promised a photo of Anna Kournakova. Instead, once again, upon opening the email attachment, a VBScript executed that sent a worm to everyone in the victim's address book. But, unlike the ILOVEYOU virus, the Anna Kournakova virus didn't damage computers or their content.

Security Updates

In response to the burgeoning number of computer viruses, which now infected millions of systems, security companies, in conjunction with the operating system vendors, began to release more sophisticated and more frequent anti-virus updates and patches. "Update, patch, update, patch" became the new battle cry.

Companies hired waves of internal IT security staff. Computer users attended 'lunch and learns' about safe email practices. VPN (Virtual Private Network) connections became de rigueur for home-based corporate laptop use. And employees were forced to sit through long computer update sessions each time they joined their corporate networks.

'*Windows Update Services- 589 security updates in this critical patch. Might as well go out for a run, and maybe dinner. Save your work*'. And this was only the beginning.

Spam

Then came the spam. And more spam. And spam flooding email inboxes around the globe in 2001/2002/2003. Spam is unsolicited messages- in this case email that hogs bandwidth, drains productivity, and is just plain annoying. Do I look like I want to buy Rogaine; I have a veritable mane of hair for goodness sakes. Oh- and spam sometimes contains malware. In fact, as spam gangs began racking up profits, they employed malware to take over third party computers, turning them into spam bot slaves. This provided free computing power and added a layer of obfuscation around spam kingpins' identities.

Enter anti-spam services and products, a new challenge and source of revenue for antispam start-ups and the who's who of antivirus vendors. While the spam problem was exploding, I was gainfully employed at a large security company working on their business gateway email security offering. A competitor released a new digital anti-spam service that was one of the first purpose specific, business-focused security services. My company's counterpunch was the first on-premise, gateway anti-spam functionality, integrated into their gateway anti-virus and content scanning suite. The race was on.

Patching and Configuration

In addition to deploying the latest security products, IT staff quickly learned the importance of keeping all of their systems patched and properly configured.

Case in point, I was visiting a Fortune 100 East Coast manufacturer the day that SQL Slammer hit. The SQL Slammer worm attacked an unpatched buffer overflow vulnerably in the Microsoft SQL Server database product. Our client was running a small number of unpatched SQL Server instances in a regional parts and inventory warehouse. The worm crashed the servers, taking warehouse and shipping functions offline for several days. The worm generated so much network traffic that several routers crashed as well.

So, in addition to reimaging the servers and patching them properly, the IT staff also had to tend to corrupted routers and routing tables. At this location alone, the attack cost millions of dollars.

Updates to Network Devices

Speaking of networking, in the typical IT department the network computing staff has traditionally been separate from the information systems staff. This means that network hardware passwords and patches may not be updated per corporate information systems policies or schedules. And integration testing may or may not occur.

The result: many an organization has been compromised by malware entering through a forgotten fax machine, printer, or an unauthorized end user hot spot. Such breaches are particularly relevant in universities, private companies, and unregulated industries, where end users both demand and have more digital freedom.

Glut of Security Attacks

The sheer glut of security attack targets is now a factor too, particularly for consumers. For example, my mother has an iMac and eagerly logs on daily for her book club chain email and political joke fix. In an attempt to keep these top secret communications private, she uses a consumer antivirus product and regularly updates her system with patches. Finally, she joined Lifelock and even password protected her computer. Sounds impressive, eh?

Well, she forgot about her router and her wireless network. By default her wireless network was enabled and not password protected (even though she exclusively uses a wired Ethernet connection). To boot, she'd never changed her router password from the vendor's default password. Heavens to mergetroid. The neighbors must have her book club's reading list by now! Seriously, remember that *everything* connected to the network needs to be password protected and patched.

The Bad Guys Always Adapt

Increasingly, the protectors hardened email, the perimeter, and the network. So, much like biologic viruses, digital viruses needed novel tricks to thwart these new defenses. Rising to the challenge, the bad guys adapted. Enter the era of keyloggers, Remote Access Trojans, and back doors. These new threats changed the game and upped the ante. Remote Access Trojans slipped pass firewalls by opening ports and communicating out of the network, in what was considered "the reverse direction".

Clever kernel level and memory viruses evaded signature-based protections. And no longer were the virus writers motivations so benign, chasing curiosity, mischief, or bragging rights. Nope. The new wave of hackers included criminals- organized and well funded. Ransom-ware was born. Hackers sabotaged or 'coopted' digital assets until blackmail demands were met. Criminal organizations turned to cybercrime. And ironically, what made the Internet such a powerful, positive tool: the breakdown of geographic and socioeconomic barriers, democratization of knowledge and opportunity, and amplification of ideas and data sharing, made it all the more dangerous.

New devices and attack targets provided novel opportunities for increasingly sophisticated hackers. Wireless networks were everywhere. A hacker could now duck into Starbucks for a juicy midday hack along with his latte. Different operating systems, browsers, and databases were on the rise. MacOS, previously a smidge of endpoint OS (Operating Systems) market share, became popular.

Dizzying Array of Technology and System Updates

New web applications were written for Internet Explorer, Firefox, Chrome and Safari browsers. Portable USB (Universal Serial Bus) storage sticks and hard drives proliferated. New identity directories and authorization schemes were deployed. Users had more choices- resulting in new functionality and reduced costs. But, at a heavy burden to IT and security professionals. This would be analogous to someone breaking into the Post Office and mixing up all of the residents names and addresses. Luckily, though an critical vulnerability existed, Dan Kaminsky found it and brought the key vendors together to patch it before it was exploited in the wild.

IT staff and developers alike now had to account for a dizzying array of security update systems, patches, and configurations. IT staff implemented new processes to package and schedule updates, which required testing in a lab environment before being pushed to production.

While working for a small software firm, I received a frantic call from a security professional at a Fortune 50 company. He was in a full blown panic because one of my software patches was in conflict with another vendor's patch. Our customer urgently needed to push out the combined update package, which contained several new features and functions. So, my small resource constrained company had to coordinate and trouble shoot with another vendor to make our patch simpatico with theirs. And note, a change to our patch meant that we had to go back and run through the testing cycle again across our entire matrix of platforms and devices.

Suddenly, security patching and maintenance was much more serious, complex and costly. Ironically, the very systems that became mission critical to security, in this case update and patching systems, in turn became juicy targets for attackers. A *Who's protecting the protectors* type of thing.

A recent example- the security defenders who reverse engineered the Stuxnet malware at first thought that the virus writers had compromised the Microsoft Windows Update Service. This would have been an epic hack with massive fallout. Luckily, this was not the case. Instead, the Stuxnet creators had spoofed the system.

Another similar close call, and this one was real, occurred in 2008 when Dan Kaminsky found a fundamental flaw in the DNS protocol. The Domain Name Services system is the addressing and routing system for the Internet. Upon exploit, bad actors could poison DNS caches allowing then to redirect network clients to their choice of DNS servers.

Common Attack Vectors

Back to the most common attack vectors. Though email remains both a common malware ingress point and spreading mechanism, the first five years of the new century saw the rise of a new entrant to this infamy: the web. In addition to browser attacks on the user side, those traveling the virtual highway encountered a variety of new threats. When one surfs the web, there is software and data on both sides of the interaction. And most of the time, the website server resides remotely, outside of the firewall and outside of local security staff control.

This opens Internet users up to the possibility of spoofed web requests, web traffic redirection, interception of the traffic, and milieus websites and web data itself.

Fake Sites

In addition web-oriented malware, websites are rich fodder for fake content. Fraudsters have been known to set up ersatz banking web sites that look nearly identical to the real bank web sites. The fraudsters then trick users to "log in" to their false sites and thus share their login credentials with the bad guys. Some fraudsters have even created websites for fake medical offices as a part of Medicaid and Medicare reimbursement scams.

The relative ease of creating fake content and spoofing website identity has driven a spate of trust services and companies that validate the identity of website owners, the safety of websites, and the risk of associated content and downloadable assets. Also, secure transmission protocols such as https have standard for secure browsing. And finally, biometrics and double factor authentication are more common than not for hardening identity checks.

Last but not least, certificate authorities sit at the apex of the digital trust chain.

Digital Certificate Authorities

Certificate authorities make it their job to validate and assure the identity of certificate owners. Digital certificates are issued for things like web browsers, web sites, and software programs. The certificate authority system is hierarchical, with a few well known root authorities anchor the system. A breach of the certificate authorities would be even more cataclysmic than a breach of the major update and patch systems. It would fracture the underlying trust of the Internet.

There have been a few breaches of smaller, peripheral certificate authorities, but fortunately the system has remained for the most part uncompromised.

As discussed, during the first decade of the twenty-first century, business servers, network attached devices and websites were typical attack targets, along with consumer laptops.

Mobile Computing

But as 2010 approached, the next digital trend caught fire- the move to truly portable computing. Large brick-like laptops fell by the wayside and notebooks, tablets, and full-featured smart phones exploded in number. What did this mean? There were a whole slew of new attack targets and attack surfaces. But, given the commensurate differences in hardware, operating systems, and applications, there was a notable lag in the associated inception of malware. Moreover, the iOS operating system had become wildly popular, and as a proprietary, closed system it was significantly more secure.

Hackers Parry, Reinvent and Adapt

Did the flood of new devices, protocols, applications, and operating systems discourage the bad actors? Not a chance. A decade into the new century, the hackers continued to do what they do best- parry, re-invent, and adapt. Case in point, the iOS XcodeGhost exploit and iCloud's celebrity photo theft. iOS, once considered immune, was no longer. And hackers became even more inventive, infamously attacking Target via their Point of Sale and HVAC (Heating, Ventilation and Air Conditioning) systems and devices. And then came the programmable logic controller (PLC) attacks. PLCs control and automate the electromechanical operation of industrial machinery.

Stuxnet was the first widely known PLC breach. Stuxnet was a sophisticated worm that infected software running centrifuges purifying plutonium for nuclear weapons. This attack is considered the first widely known act of a national-state engaging in cyberwar.

We Can Do Better

Wow. Malware writers and those intent on digital mischief seem to have the run of the landscape. Well, not necessarily.

I stated at the beginning of this article that I wasn't all doom and gloom when it came to security in the twenty-first century. How can this be given what I've told you. Given the huge increase in targets, the wide variety of targets and attack surfaces, and the sophistication and connectedness of malware writers and the hacker community at large. Computer antivirus software wasn't enough. Protecting the perimeter wasn't enough. Protecting the network hasn't saved the day. And more people, machines, and data are spoofed, sabotaged and stolen than ever. What can we do?

Rest assured, we can do better. First of all, we need to have realistic expectations. Anything connected to any network is at some risk. Absolute security is unattainable. Given that baseline understanding, consumers and corporate security defenders must critically assess digital assets, data, and behaviors, creating a value to risk matrix. Obviously, assets and information with the highest value and highest risk warrant the most focus and the most security spend.

Next, three key factors I mentioned earlier come into play: advanced computation and analysis of disparate pools of data; elevated awareness as security becomes increasingly strategic; and the continued acceleration of innovation.

First of all, let's examine the astounding advances in data analysis and computation. Let's examine these three factors in order.

Data Analysis and Computation

Why is this so important? Simple- the volume and variety of security information. Remember, security products, services, and platforms receive endpoint data, network data, website data, mobile data, application data, wireless data, and hardware data. And, it's coming from everywhere, all the time. Early products called security information managers attempted to collect and normalize a wide array of security information and send back actionable changes and directives. But they were hampered by processing requirements, the elementary nature of algorithms, and the difficulty in normalizing disparate data.

The march of innovation, on all fronts, is addressing early problems. We continue to enjoy mind-boggling leaps in computer processing. Hardware and software advances, along with next-generation ASICs (Application-Specific Integrated Circuits) and even bleeding-edge disruptors such as quantum computers, have enabled organizations around the world to analyze their internal, external, and supply chain data through different lenses, more completely and deeply than ever before. With the staggering volume of continuous security-oriented flows, experts have made strides in data modeling, abstraction, and transformation.

All of these improvements allow different data: such as geospacial, audio, industrial control, and even complex video data to be consumed and processed right along with text, image, and meta data. Imagine the possibilities. New high-speed cameras are able to capture biochemical and physical reactions at the atomic level. Improvements in artificial vision can resolve images better than the human eye. The latest satellites can detect dynamic events in space and on earth with exponential precision. Hence, new algorithms will assess such video feeds against audio recordings and other event captures and knowledge, improving deductive and predictive accuracy at unprecedented levels.

As service providers, governments, corporations and consumer collect and track cleaner and more insightful security data, actions, and closed loop feedback, the information stores increase in value. New ways of sharing such information, while protecting confidentiality, foster novel collective insights and knowledge. Through advances in deep learning and artificial intelligence, these insights and knowledge will only improve.

Deep Learning and Artificial Intelligence

Deep learning and artificial intelligence are prime beneficiaries of increased computing power because deep learning requires huge quantities of data and the ability to run thousands of simultaneous calculations and simulations to ferret out results that often run counter to intuition and common sense. Security professionals will increasingly rely on deep learning and artificial intelligence given the flood of security data, incidents, and complex workflows that have become the status quo. For instance, through deep learning, a security professional might discover that contractors hired via a specific vendor, who created patches for mobile networking codebases for the company's Hong Kong geography, left back doors in the code that were exploited to steal intellectual property. Discovering this causal link could be nearly impossible via manual human inspection, particularly given the number of systems, databases, and touch points the causal chain crossed.

Deep learning and artificial intelligence insights will allow security teams to 'find the needles in the haystack', to be sure. But they will also empower security teams to apply contextually relevant automation and remediation, real-time, while highlighting well-described critical issues. In addition, security professionals will be able to deploy just-in-time alerts, warnings and prompts that can tease out malicious acts from the unintended and accidental. For example, a warning might tell users that they are attempting to access top secret materials during off hours from a VPN ingress point- and request acknowledgment of this fact. Thus, users who didn't mean to take such actions can stop before they either break rules or take undue risks. Finally, deep learning and artificial intelligence will enable analysis of security sensor data, events and incidents against real-time financial information, production data, updates in laws and regulations, market and competitor news, and even brand equity measures-permitting more informed cross disciplinary executive decision making. It will be easier for business leaders to weigh metrics-driven risk against innovation and increased revenue, both cross-functionally and by product, market, geography, or buyer profile. Over time, automation, remediation, and "just in time" user behavior modification will be built into products from the ground up. This will harken the age of "security as a service", innately. Security will be cheaper, stronger, and integral. Thus, deep learning and artificial intelligence, appropriately applied, should be embraced.

Summary

In summary, the wild west of security has curled many security and business leaders toes for the past thirty years or so. We braved the advent of security attacks by user profile and characteristic, access point and method, device and asset type, operating system, language and currency format, social engineering, identity and verification method, etc. And then the attacks became multi-pronged and multi-layered. And the number of devices with programmable logic chips and Internet access exploded. Finally, the number of digital users blossomed, growing ever savvy and connected.

Eek. But, as I claimed, three key forces will help turn the tide, elucidating the digital security landscape and helping tame it. First, security awareness and quality education will become more pervasive. Social engineering and cooperative online efforts are already speeding this along. Concomitantly, vast improvements in computing power, memory, connectedness, flexible architectures, and smart sensors will foster deep learning and artificial intelligence solutions that will help tip the security balance towards the defenders. We will realize fewer costly breaches along with smarter overall business decision making. Viva Les Information Technology and Security Superheroes!

Chapter 7
Hidden Dangers of Internet of Things

Martha Daniel

Overview

Computer security has evolved over the past 25 years as innovative technologies embrace wireless telecommunication integrating information and data and connecting the world globally. I can remember in the early 1990s the announcement that the "Super Highway" was coming with the prediction that it was going to significantly change the way we do business, the way we live, and the way we work.

Now today over 25 years later the "Superhighway" opened avenues and roads thrusting information and connecting our world globally. In reality, this revolutionary technology created a "new world" a domain we call "the Internet". Within this "new world", we have been existing without global policies, rules or regulations, treaties to guide collaborations and protection. Every country has embraced this "new world" and within each country they have declared jurisdictions and territories which are hosting and housing public, private and federal information and data.

The increased volume of information and data being introduced into this "new world" also introduces serious demands for protection and computer security. Today, we are challenged with protecting this information from those who desire to take it and capitalize upon it through theft, corruption, and fraud. The goodness of this innovation far outweighs it negatives; however over the past years the "bad guys and gals" have increased their presence which has now required us to focus on computer security and protection which today is called, Cybersecurity.

The worldwide problem we face today with the internet is—how do we keep the "bad guys and gals, other countries, and terrorist from taking the information and data from this "new world" to specifically corrupt and destroy the world we live in today.

M. Daniel (✉)

D.A. Christofferson (ed.), *Women in Security*, Women in Engineering and Science, https://doi.org/10.1007/978-3-319-57795-1_7

The increasing number of devices that are interconnecting us worldwide and more and more innovation of the utilization of these devices in a wireless environment is driving the need for Cybersecurity and governance. Unethical practices, malicious attacks on data, and corruption have already started Cyber wars. There are so many unknowns, threats, and vulnerabilities associated with the internet today and yet we are now moving towards an advanced age of information now called "The Internet of Things—IoT".

We have so much more work to do with "the Internet" and now with "the Internet of Things-IoT" we are challenged with more hidden unknowns and uncertainties that will surely impact the world we now live. My company, IMRI has been a significant player in the technology industry through the dynamic evolution of technology since 1992, and we are no strangers to Cybersecurity. Our Cybersecurity Division, Cytellix, www.cytellix.com, has been engaged in the protection of the internet worldwide now since 2007 and currently offers a Managed Services Subscription service offering that protects information and data providing real time continuous monitoring of networks bringing complete visibility of all assets and devices residing on your networks, and bringing unique intelligence from threat feeds which identifies and profiles "bad guys and gals—any intruders" who attempt to maliciously attack your company's public or private information. Whether the company is a large entity or small business, we have proven, tested, operational, and affordable services and solutions that bring winning results and outcomes in the war against Cyber intrusions.

This chapter will focus on the "Hidden Unknowns of the Internet of Things". After reading it, you will better understand the challenges that our world now faces as we expand beyond "the internet" to now, "the Internet of Things".

What Really Is the Internet of Things: IoT?

Internet of Things, IoT, refers to internet-connected devices which can interact with other connected devices or objects over a network. IoT is a network of physical and sometimes virtual platforms with IP (Internet Protocol) addresses to enable internet connectivity. These platforms use built-in technology to interact and communicate with other Internet-enabled systems. Besides common computing devices such as laptops, desktops, tablets and smartphones; you can connect other objects with IP (Internet Protocol) addresses to the Internet.

Household appliances, baby monitors, smart locks, smart TVs, wearables, utility devices and other devices with built-in computing systems can be connected to the internet. Electronic appliances, alarm clocks, speaker systems, security systems, thermostats, vehicles, light bulbs, refrigerators, pacemakers and others can have computing systems embedded in them. These devices are known as smart or connected devices. Each device or platform has an identifier, IP address, that can allow specific objects to transfer data through the internet.

The IoT is presently a high volume emerging technology in today's world. According to Gartner, by 2020, suppliers of IoT products and services will generate a revenue increase of over $300 billion. They also expect about 20.8 billion IoT devices. This excludes tablets, smartphones, and computers. IDC, on the other hand, predict a $7.1 trillion growth from $1.9 trillion in the worldwide market regarding IoT solutions in 2020. They expect 28.1 billion devices excluding tablets, smartphones, and computers. IHS Markit professionals predict about 30.7 billion IoT devices by 2020. Regardless of which estimate is on target, we anticipate over 20 billion interconnected devices in the next 3 years.

What Are the Benefits and Significance of IoT?

There are several benefits attached to the revolution of the IoT. These benefits will on daily basis help individuals, businesses, and society as a whole. For individuals, this new concept comes in a lot of forms ranging from safety, health, and everyday automation.

The introduction of IoT into the health care system has proven to be greatly beneficial to both individuals and the society. Hospitals are able to monitor patients vital signs by implementing a chip into individual monitoring devices. These chips provide information to help doctors to determine whether or not a total assessment of the patient is needed. With hospitals struggling on a daily basis to take care of the vast number of patients they have, monitoring each patient's health will allow them to decide who needs attention first. There is a significant cost savings to our healthcare system when we can use automated monitoring as part of preventative techniques.

The Internet of Things can also assist people with their personal safety. ADT, Comcast, AT&T and others provide home security systems that allow individuals to use their phones to monitor and control their home security.

GM's OnStar, Ford's Sync and Chrysler/Dodge's UConnect technologies which are placed in vehicles, provide navigation assistance; allow remote starting and remote locking or unlocking of doors; and detect when a crash occurs and calls 911 automatically. They can also efficiently track any motion of vehicles.

IoT creates an avenue that can help in saving money for people within their homes provided their home appliances can communicate. A smart home is a perfect example of interaction between household appliances and other devices at home. Appliances can be operated in a way whereby energy usage is efficient. The home can provide comfort and convenience based upon the lifestyle of the owners. The garage door opens once the car communicates with it. The homeowner's smartphone unlocks the doors, and the intensity of the home's lighting is set as the user desires. The temperature of the thermostat is adjusted to suit the user's choice. The devices in this home interact with each other to improve the user's lifestyle.

Businesses can equally gain a lot of benefits from the Internet of Things. It can be employed in several different categories which include inventory control, asset tracking, security, shipping and location, energy conservation, and inventory tracking. Devices are able to communicate their status, location and other pertinent information.

Another benefit of IoT is its ability to track individual consumer behaviors for marketing programs. Each customer is targeted based on the information provided by the device. Marketing programs that use consumer behaviors have been proven to increase business sales and knowledge of demographic trends.

Security Risks of the IoT

The prevalence of objects with digital identifiers that allow them to interact with their environment through the web has become a great concern to security experts. These connected devices are beneficial, but most of them have exploitable vulnerabilities. These devices are not secured, and access to them is rarely restricted. The computing systems embedded in these devices are typically not secured as the design considerations for protection of these objects was not considered. As a result of the insecurity, traffic and access to the smart devices cannot be controlled. This poses a global security risk.

Cyber criminals and unethical hackers can control these smart devices with no more than an internet scan for a manufacturer identifier. Through a basic scan of the internet, a hacker can find a significant distribution of a simple set of devices, such as web cams. With the location identified and the device security flaws understood, the hacker can perform a targeted broad-based attack of the device owner's networks. Recent attacks on the Dyn DNS (Domain Name System) servers, which brought down Amazon, Twitter and Netflix, originated from Mirai botnet infected cameras. Hackers took advantage of IP-based cameras by knowing the device settings and their inherent weaknesses and used tens of millions of smart home devices connected to the internet as weapons in the cyberattack against Dyn.

Users of interconnected devices must be vigilant as long as their devices are connected to the internet. They should be aware of the risks involved in using such devices. Some examples prone to risks include; home security systems, connected cars, smart fridges, smart locks, and smart light bulbs; yes, light bulbs.

The use of smart devices and systems to secure homes has created new security and safety concerns. Smart devices can compromise your privacy. Criminals can take advantage of the weak security of these devices to attack your home network. They can access smart devices, monitor your activities or steal personal data. With the popularity in the use of IoT devices, exposure and risk of cyber attacks on smart homes are on the rise. Most recently, several popular smart home devices including *Nest, Ring, SimpliSafe,* and *SmartThings* were shown to have security flaws that potentially exposed homeowners to all sorts of problems. Also, researchers at the University of Michigan were able to hack into another leading smart home automation system and get the PIN code for a home's front door.

Many users are not aware that in addition to the IoT devices, routers can also be exploited. The router connects devices in your home or business. Flaws in routers can make them vulnerable to exploitation by cyber criminals. They can interfere with the functioning, data and network in your home or office once they can control the router.

Car hacking has been on the increase. In 2015, Chrysler announced a recall for 1.4 million vehicles after two researchers demonstrated that they could remotely hijack a Jeep's digital systems over the Internet. While the recall cost Chrysler a lot of money, if those two researchers had exploited the vulnerabilities the way malicious hackers would have done instead of reporting their research to Chrysler, things could have been much worse.

Some devices can be hacked and compromise the safety of users, not just the building they're housed within. Traffic light controls, healthcare services, and industrial systems are at high risks and can cause havoc to safety and emergency response. For example, in October 2016 hundreds of operations, outpatient appointments, and diagnostic procedures were canceled at multiple hospitals in Lincolnshire, England, after a malware hack compromised the National Health Service (NHS) network.

What Can You Do to Protect Your IoT?

You cannot predict cyber attacks, but you can secure your smart devices against them and reduce damages resulting from these attacks. Protecting and managing IoT devices can be a major challenge, especially if you don't know everything that is connected to your networks. You should adopt a situational awareness model of all devices on your network so you have clear knowledge of every asset on the network, their security posture and their communication paths. Knowing about all devices and their cyber posture will help prevent cyber-crimes in the future.

Early detection of threats to connected devices by monitoring in real-time is crucial in the mitigation of security risk posed by these devices. Adoption of a system that prevents intrusion by shutting down the network whenever anomalous traffic to these devices is detected will be helpful. Automated alerts to new devices and behavior changes can inform when criminals may be attacking the network. This monitoring the smart devices environment will help you decide on the best approach to handle an attack and lessen damages if an attack occurs.

Cyber criminals have frustrated owners of smart devices by taking advantage of devices that are using factory default usernames and passwords which make it easy to gain control of these devices. It's important to reset default passwords to be unique and complex.

Security experts have created cryptographic algorithms to help protect identities and control access to sensitive or personal data. Ensure that interactions between interconnected devices are encrypted. Use encryption to store privacy data including your Wi-Fi password. The use of encryption will protect your connected devices if a hacker gains access to any of the devices.

Regular update of firmware of embedded computer systems and applications can help secure a smart device. Verify that updates are from the OEM (Original Equipment Manufacturer) or developer before applying them.

Conclusion

The Internet of Things no doubt provides several benefits to individuals, business, consumers and ultimately to the society. I am hopeful, that as the evolution of the products and services evolve, the improvements in security will also be seen.

About IMRI

IMRI is an industry-leading provider of cybersecurity, technology, program management, and engineering services for government organizations and commercial enterprises. IMRI is a change integrator that leverages its 25 years of highly-specialized data center expertise to provide its clients with integrated, solution-based programs to help them meet the requirements and challenges of an ever-changing business environment. Working with some of the largest organizations and networks in the world, IMRI operates in 19 states, providing its clients a critical combination of corporate experience, localized expertise, and proven methodologies and tools that integrate seamlessly into any enterprise.

About Cytellix

Cytellix is a privately held cybersecurity consulting firm specializing in comprehensive network intelligence and situational awareness. Powered by Enterprise Situational Intelligence (ESI), Cytellix has the only solution in the industry that can detect known and "unknown" threats in any environment, while providing complete network visibility. A division of Information Management Resources, Inc. (IMRI), Cytellix manages millions of IP addresses for organizations of every size in a wide range of data-rich industries—including government, manufacturing, finance, banking, law, education, healthcare and municipalities—with best-in-class, real-time network scanning technology. In addition to securing network perimeters for the U.S. Army and the Missile Defense Agency, as well as leading corporations such as PricewaterhouseCoopers (PwC), Kaiser Permanente, and the Walt Disney Company, its agentless, scalable solutions are utilized by small- and mid-size companies for compliance assessments, one-time audits and in preparation for mergers and acquisitions. With a goal to drive business growth by protecting and defending critical enterprise IT infrastructures throughout the world, Cytellix is well on its way to revolutionizing the cybersecurity industry.

Chapter 8
Information Security: Beyond the Bits and Bytes

Mary Ann Davidson

Information security is many things, but boring is definitely not one of them. It's a rollercoaster where the ride gets more dramatic every year, zipping down a plunging chute and taking your breath away. I've worked in information security for most of my professional career. I got into it when I became bored with building accounting software, and there was an opening in a group at my company that was building a product to "automagically" handle data of different security classifications (e.g., people who can only access "Unclassified" information wouldn't be able to view or change information labeled "Secret"). As a former U.S. Navy officer, I knew something about the value of limiting data access based on security classifications. I took a job as a product manager in that team and almost immediately found what "tripped my trigger" in the tech sector: security! Since that first security job, I've worked with a bunch of different teams and technologies and enjoyed (almost) every minute of it.

Working in information security is different from many other aspects of technology in that there are larger issues than the bits and bytes. For example, if you are stranded in the middle of Upper Slobbovia, a map application that knows *exactly* where you are might help you find the fastest route home, even taking traffic and weather conditions into account. Neat, huh? Of course, it is not so neat if someone is able to use those same geolocation services to track your every movement, unbeknownst to you. The same technology that can be used to make lives easier and more productive may also have "unintended consequences." Information security involves actively looking for those "unintended consequences." As I learned from military history, "where there is capability, an enemy may develop intent."

M.A. Davidson (✉)

D.A. Christofferson (ed.), *Women in Security*, Women in Engineering and Science, https://doi.org/10.1007/978-3-319-57795-1_8

Security Is a Team Sport

You only as good as the people willing to work with you and for you. That is, the old security saw about "a chain is only as strong as its weakest link" is even truer when it comes to building a strong security team. You want really good people with a variety of skills and—equally important—a variety of *viewpoints.* Unless you believe you are like Mary Poppins— "practically perfect in every way"—it's really important to have people working for you (and with you) who have a diversity of viewpoints and the freedom to express them. You will make better decisions that way: I know I do.

My team encompasses people with a variety of skills. Ethical hackers try to break our products and services before bad guys do; vulnerability handlers help developers find, triage and fix security vulnerabilities; program managers implement the assurance program by which we "build in" security across the breadth of our products and service offerings; security evaluators certify our products against relevant U.S. and international security standards (such as the U.S. Federal Information Processing Standards (FIPS) 140, to validate our encryption implementations). We work also with other security experts across the company in a variety of areas: legal, public policy, compliance and operational security. My job in information security is many things, but it is never boring.

Soft Skills Matter

Even if you are a technical whiz like the ethical hackers who work for me, you still need to hone other skills, which can be equally important in your ability to be an effective information security professional. One if the more important skills is—communication. Yes, I know, not a "sexy" skill, but still very important. For example, the real job of ethical hackers on my team isn't "finding security problems," it is explaining *what* they found, *how* they found it, in a way that the larger lessons are captured and mere mortals can understand and act upon the information. If you start a discussion with "here is how we trashed your code," the people who wrote the code will get very defensive and not listen to you. If you start with, "we are here to help you by finding problems, explaining them, and giving you a chance to address them," you will get more "takers." De-geeking the speak is important, too. If you say "there is a clearly a BBSP in the frabistat object leading to a CST," you might as well be speaking Homeric Greek: you won't be very well understood. (Admittedly, I made up those acronyms.)

Two of the soft skills that have worked well for me are the ability to use analogies and to analyze problems in economic terms. Analogies are important because they help people understand something by relating an unknown to something they know: "this is like that." I create analogies when someone is explaining a technical problem to me, to help ensure I've understood the problem correctly. Sometimes I use an analogy to illustrate a larger truth. For example, there is a huge gap between the industry need for cybersecurity experts and the number of people who can fill those slots. "We need more cybersecurity experts!" is a true statement but it ignores what I think is the bigger problem.

Remember the story of the little Dutch boy, who, detecting a leak in the dike that surrounded his town, bravely put his fingers in the hole to prevent a flood. Too many people think the answer to cybersecurity threats is more little Dutch boys—tens of thousands of them—to plug all those holes! The real problem is a failure of engineering, the result of which is that sooner or later, all those dikes will break and we will have 30,000 drowned Dutch boys. In short, we need to engineer our systems to be more secure and more attack resistant, or we will never be able to hire enough "Dutch boy cyber defenders" to secure them.

Economics rules the world, because there are some fundamental truths that affect what you do and how you do it, no matter what business you are in. One of them is that resources—time, money and qualified people—are always limited. Sure, you may be able to hire more people—if you can find them—but you likely will never have enough time, money *and* people to do absolutely everything you want, all at the same time, to perfection. That is all the more reason to look at what we do from the perspective of what matters the most, and what we can do that is highly *leverageable*. My team has embedded security requirements and checklists into the tools development organizations already use to track milestones (better to fill out one checklist than two checklists—it's faster and easier). We use metrics to enable development teams to manage their own workload better ("here are the most important things to do first"), which helps teams use their resources better. Cost avoidance is another economic way of thinking about security, especially, the business case for finding and fixing security issues earlier. It's better to fix something earlier, once, than have to fix it later—and more expensively—across X supported versions and Y operating systems. Economics is your friend, especially when it comes to "doing the most security good in the most efficient way."

Change Is a Constant

Mechanical door locks haven't changed all that much in eons. There are tumblers, there's a key: the fundamental mechanisms are pretty well understood. Most of what we deal with in the information security world isn't so static. (Not even what we call it: we've morphed from "computer security" to "information security" to "cybersecurity.") Information security changes rapidly because information *technology* is changing rapidly. There are some fundamental security truths (rule 1 is "never trust any unverified input" and rule 2 is "see rule 1"). However, the technology changes, where we use it changes, and how we interact with it changes, even if the threats just get exploited in new venues and new protocols. In short, you need to be adaptable if you want to work in information security.

My team has adapted what we do to the increased delivery of information technology as cloud services rather than on-premises products. The move to cloud in no way diminishes the importance of security and provides new opportunities to have better security at lower cost. For example, when you deliver a product, you don't always know exactly how customers will use it in their systems. You should—and most vendors do—deliver a secure default configuration so a customer is reasonably secure "out of the box," but you can't know everything about all customers' security needs. Similarly, when you drive a new car off the lot, the seatbelts are already configured and the air bags are ready to deploy, but the manufacturer cannot know how much leg room the driver needs, so the driver must adjust the seat to her liking.

With cloud offerings, if you are the one running them, you have a much better idea exactly how they are going to be used, which makes it easier to lock them down ahead of delivery, increasing security "out-of-the-box," and at a lower lifecycle cost. Even better, you can do things that lead to lower cost of secure operations (like having specific audit options on by default): stronger, repeatable, "operationalized" security. I enjoy watching technology change and ensuring that what my team does continually adapts and improves security.

Embracing change also means that you reexamine your beliefs and your practices from time to time. Business changes, customer expectations change, threats change, and that means that you need to ask questions like, "we've been handling Y a particular way and for what were good reasons at the time we created our policy. Is that still the right thing to do?" Never say never, and be willing to reexamine what you do and why you are doing it.

Integrity Is key

Integrity in the information security context means that data has not been altered inappropriately: that is, you have confidence that "A" is "A" and not "B" that has been futzed with—inappropriately altered or corrupted. That kind of integrity is really important because data you cannot trust is probably worse than no data at all.

There is another kind of integrity that is even more crucial in information security: doing the right thing, including giving your best professional opinion. In security, you may frequently be explaining risks or potential threats to people who are not as geeky as you and, to be fair, have different and perhaps broader responsibilities. Integrity includes giving one's best professional opinion regardless of the circumstances, rather than "telling Ms. X or Mr. Y what you think she or he wants to hear." Of course, part of being a professional is not just describing a problem but providing context (what is the ramification of this?) and ideally a recommendation as to how to address the problem you raised. It also means that, having given your best professional opinion, you implement whatever management decides to do to the very best of your ability.

One of the best parts of my job is that I feel I have always been supported for giving my best professional opinion—even better, I have been *valued* for that. On occasion, a manager may have made a different business decision than I recommended, but the difference has never been one of principle, just a different (yet still ethical) choice. Ultimately, your work environment is perhaps the key factor to success in information security, because you can only effect needed change if you have the ability and the opportunity to "let your 'yes' be 'yes' and your 'no' be 'no.'" When you work in an area as challenging as information security, with a great team, and your core integrity is valued and respected, it's a terrific place to be.

Chapter 9
Building the Bridges to Security

Miriam Fernandez

Overview

Over the years, I have solely resided in IT (Information Technology) organizations, until now, where I work in a more evolutionary function. My achievements have directly supported the businesses that depend on IT to run. Core competencies include engineering and administration of the network backbone, plus UNIX servers supporting the IT Infrastructure. These directly connect the IT environment to the business groups and functions.

In addition to technology, my roles have evolved and created policies and procedures, standards, and process improvement. These strengthen the IT platform that the business runs and depends on.

IT server and network expertise create a natural bridge to technical competence in security. My transition from IT to security was gradual, as my security responsibilities increased over time. In different positions, the confluence of the roads all led to security—or Rome!

Although responsibilities along my career path have evolved, they always fell within IT roles and included:

- Compliance
- Security policies and procedures
- Network and server infrastructure support
- Disaster recovery
- Physical security in IT

My current security position is with Intel Corporation, in a role supporting the Intel Key Generation Facility. When interviewing for this position, the experience that stood out to the hiring manager, was my background in business continuity and

M. Fernandez (✉)

D.A. Christofferson (ed.), *Women in Security*, Women in Engineering and Science, https://doi.org/10.1007/978-3-319-57795-1_9

disaster recovery, and security compliance. My network infrastructure and IT operations background created a foundation, but the noted experiences created the standouts on why Intel took a second look.

Past: Network, Security and Compliance Improvements

In my first exposure to security: I worked in an IT network group, in a technology company with 300,000 global employees, within an enterprise services team supporting the Americas region.

My security start came from UNIX engineering and network support, in a UNIX System Administrator role. Within this IT network support role:

- Requests came in to scan servers to see if they were configured securely to protect the company's network and data
- Servers were scanned, reports were generated, and report results were reviewed
- High and medium priority findings had to be addressed
- Findings were evaluated first, for impact to the environment:
  - Patching could break something or cause downtime
  - Some legacy systems did not have patches—and required other mitigation options to be evaluated
  - Problem solving skills were required, to work out which steps to take
- When a finding could not be resolved, an "Exception" was created (security policy exception process)
- To comprehend, document and get this approved, meant working with colleagues, business units, and different groups of people in security for completing the cycle. Multiple groups of people within the information security team were involved.

The Archer tool was used to create policy exceptions in the system and get them approved. Archer is a Risk Management Database. At this point, I was doing less server administration at HP and doing more security compliance.

I had earlier responsibility in IT Operations to inventory all IT assets—hardware and software systems, and software licenses, including tracking for expiration renewals. This supported capital spending, expenses, and hardware depreciation for the IT Data Center.

Financially, I supported project scoping to buy systems, hardware, and software solutions. Included: Collecting requirements and quotes, and submitting to IT team for justification and prioritization. We looked for solutions that worked, then bought and implemented them.

In HP's Transformation Team, I shifted more towards a DR (Disaster Recovery) role.

Disaster Recovery Zero to Hero

My role at HP was UNIX System Administrator and the UNIX/LINUX servers I supported were transitioned into virtual servers. During this time, I was assigned the role of Disaster Recovery (DR) Team Lead for the Americas team. That required me to coordinate with people all over US states and Canada for meetings and travel arrangements to support DR.

Early on, disaster recovery mostly consisted of data backups saved to magnetic media, which were sent offsite for storage. They were not necessarily tested to see if they worked on a real recovery.

Within the HP DR team, we built out the DR plan. This required collaboration, and getting all the IT folks—administrators, network engineers, all the people from IT along with all the business units, to identify someone from business unit to take ownership (product line). We worked with all these people to build out the DR process.

We then went offsite and tested it. We built the servers from scratch, created backup disks and built up the network starting with the infrastructure. Business units were asked to test the systems once they were up and operational, to make sure that functionality worked correctly.

It was hard to see the light at the end of the tunnel. It took a long time and it paid off.

DR is often related to security, but they are not often together. DR is a niche that is often under-appreciated.

Unmanned Telecom Field Offices

Problems in the field existed at unmanned field locations where no one was onsite. Someone only showed up if there was a problem and we had to go onsite to figure it out. We needed a mechanism to send error messages back, for remote trouble shooting.

We also had a challenge in how changes were made at the site for telecom switches. If a change was required, a staff member had to go out to the field office, and make manual changes to the master disk, at every site.

We researched solutions with an architect and engineering team in research and product development, to make this more efficient.

We arranged vendor demonstrations, and selected a product that allowed fast disk duplication and site shipment. We created a standard build, for software updates to telecommunication switches. It saved the company a lot of money and made the business run faster.

Overhauling the IT "Help Desk" for the Data Center

Early on, the IT support desk was a generic call center—the staff answered the call and took a message. It was not very helpful and added little value.

As a result, I was asked by management to improve the process. I took on the project. It gave me the opportunity to change the entire help desk environment for the IT Data Center. Workspace layout was redesigned for what would work better, and all the processes overhauled.

To prepare for the role, IBM provided training to administer the IBM mainframe in Operations. We learned how to look up problems in the system documentation—take the error code and find out what it meant.

There were no more 30-seconds to answer the call and get to the next one. Instead, the team was now working through the call, leaving the Command Center, going to the office and sitting with the right person to work through the problem.

Our team trained with second level support. We learned and gained greater trouble shooting skills to resolve and close more problems on the front end.

We went from being a call center to resolving 98% of calls on the first call, without having to escalate or ask for help. Everyone's skills were raised. Second level staff members were teaching and shadowing us, to raise the skill levels significantly.

We also researched software packages. We invited vendors onsite to demonstrate software for knowledge management systems. We wanted to enter call details online and complete a quick hit/search to see if the problem has been seen previously. "Here is what happened and here is how they fixed it." The problem might also be different but share commonalities.

We built up the knowledge management base. This is still important today, for security front line trouble shooting. It allows quicker triage, mitigation and resolution. This was my first opportunity to take something from scratch and improve it.

Raw Cable to Wireless Hosted Networks

The network has evolved enormously. Telecommunications and IT had no desktop PCs (Personal Computers) when I started in the field. We had to run the network through the wiring closet and all the way though the building to connect everybody's work station. Wire was run all the way through the building infrastructure from the floor and walls to the offices and computers, and devices configured to make it all work. It took a lot of manpower to support the telecommunications infrastructure and network that supported IT.

This evolved from raw cable to more complex networks, and to the wireless, mobile, hosted and cloud environment you see today.

Present: Protecting Devices, Content and Transactions over the Internet

Today, protecting data using strong encryption is a significant role in industry, and the one I hold at Intel. The group generates encryption capability for technology we are using in the world and going forward.

My role supports the Intel Key Generation Facility with a working title of Information Security Specialist. This is outside of IT and part of a content platform security group for software solutions. Multiple security groups exist.

This encryption capability is becoming a standard capability. It authenticates devices to the network, secures content, and safeguards financial transactions. In today's connected world, and with the Internet of Things (IoT), this is critical to properly identify devices connecting to your network and systems.

EPID Background (Source: Wikipedia)

EPID—Enhanced Privacy ID—is the cryptographic capability built into the hardware platform, for authenticating a device and providing optional content and privacy protection. While the technology already exists, Intel intends it to become an industry standard to authenticate devices for IoT. EPID provides IoT security and privacy for processors within the device sensors.

This is a summary of a complex technology capability that is not meant to be inclusive.

The EPID is Intel Corporation's recommended algorithm to attest that a trusted system is preserving privacy. It meets the Trusted Computing Group (TCG) and ISO/IEC international standard for authentication.

Originally introduced in 1999 to identify end points, the capability was removed when privacy advocates raised objections. Intel returned it to their chipsets in 2008 and to their processors in 2011.

The EPID intellectual property was contributed by Intel to ISO/IEC under RAND-Z terms. Intel also licenses EPID to third-party chip makers. Intel is recommending that EPID technology becomes the industry standard for authenticating devices on the Internet of Things (IoT). It allows inherent key distribution with the processor chip, with optional privacy benefits.

EPID is an enhancement of the DAA digital signature algorithm that supports anonymity, though the use of public verification keys and private signature keys. It allows a device to prove to an external party what kind of device it is, and optionally what software is running on the device, without having to provide device identity (authentication to a group). EPID provides a utility also, that can revoke a private key, based on a signature created by the key, without the key itself being known.

EPID is used to:

- Attest applications within the platforms used for protected content streaming and financial transactions
- Provide proof that a part or device is genuine, and that the EPID key was not revoked, using cryptographic capability within the key
- Provide content protection and attestation for secure streaming of DRM—Digital Rights Management
- Secure financial transactions using DPT—Data Protection Technology for Transactions for 2-way authentication for a POS (Point of Sale) terminal to a backend server
- Securely provision cryptographic key material over the air or down the wire
- Attest and secure the IoT: Provide authentication and preserve privacy
- When placed in devices: Provision other keys for other services in a device
- Disallow users to be tracked by IoT devices while using those services
- If required, the choice can be made to make it known (Example: For a financial transaction)
- Provide persistent identity and anonymity

Building the Tool Kit

With Intel, I've been in my role supporting this technology, for over a year now. When hiring on, I moved from one Fortune 500 technology company to another. Ninety-five percent of the job I could already do, and other parts, I found the opportunity to build skills. That attracted me—the opportunity to grow and add new skills to my tool kit. That is what we do through-out our career—build our tool kit.

On cryptography, I thought I had a good understanding, but now I know so much more. I get to work with the development team, and they explain a lot about why things are architected the way they are. This was all new territory. I learned a lot about the team and what each one does within the team.

Business Continuity Planning/Disaster Recovery (BCP/DR)

BCP/DR—on this team, I learned much more about what it takes to keep the business going in the event of a real disaster.

BCP helps keep things going, not just for disaster, but for everyday things you may not think about. It covers all aspects of the workplace—people, data, technology and processes. Anything new that is introduced to environment, you must review and assess it: What is being taken on, what the risks are, and how it will be maintained.

The BCP outcome is to recreate what exists in the environment, if it is lost or an outage occurs. The goal is to keep the business running—sustain the business.

Testing is important. The goal is to shut down the capability, bring it up at another site, and run production for a set period without issue. Work must be precise without customers being aware and with no impact.

Six Sigma Process Improvements

Since I was not in manufacturing, Six Sigma never came up for me before. I earned my Six Sigma Green Belt at Intel. It is about looking at a process, to eliminate, automate and innovate. Look at the process to see what you can eliminate—that is not necessary, what you can automate, and where you can innovate to make it better. This is internal to Intel but outside of my security group. At Intel, there is a required annual cost savings to implement a project. You have to be sponsored by someone who already has their Six Sigma Black Belt. Intel has a list of opportunities, one came up for my group and it was assigned to me. Going forward, you must do a project every year to maintain the Six Sigma Green Belt.

It has been a good year and I've learned a lot.

Future: Professional Development

Continuous learning and growth are required in any field, but especially IT. Technology is evolutionary and you are always learning on the job. The environment changes like a kaleidoscope. You may be unaware of just how much is being shaped or what you are helping create at any given moment in time.

Education

Education is important. For a technology career and success today, an IT related degree is required.

When I was in IT, the company required everyone to have a degree, if they wanted to keep their position. At the time, no one had a degree. Tuition was reimbursed for a grade of C or better. To advance, I went to school and earned a Bachelor's Degree in Information Systems. Timing was fortuitous and beneficial.

Engagement in the Field

Engagement in the field, such as a board role in a technology association will help shape you more in the external world. Leadership opportunities will grow you in new ways and expose you to new technology, people, companies and events that would not be in front of you otherwise. They also help keep you current on technology, and introduce you to people who are facing the same security challenges that you are.

Technology groups I have supported in a board role: ISSA—the Information Systems Security Association's Phoenix Chapter, the SDSUG—Sonoran Desert Security User Group, and currently I serve the CSA-SW—Cloud Security Alliance Southwest Chapter.

Conclusions

Security is an evolution that grows from the foundation we build. It happens over time, never through a single action or achievement. Security's shape shifts with the environment and our own part within.

IT engineering and operations present a solid starting or transition point for those wanting to contribute to security innovation. Continuous improvement and ongoing growth are requirements to survive, and an IT degree is needed to land most any role today.

The field moves fast, and you will have to run to keep up.

Opportunity abounds for career choice, where you will be making a difference for the future.

Chapter 10
It's All About the Cocoa Beans

Pamela Fusco

A Pioneer in the Cybersecurity Field

Navy Cryptologist

Starting as a cyber defense strategy, I served in the U.S. Navy as a Cryptologist. Spending 11 years at a global level, I worked special missions for National Security Agency (NSA) and the Navy. This created the foundation that got me to where I am today, in the cybersecurity field. Following Military service, I completed a stint at the White House doing Special Operations.

Leaving the White House assignment, I moved to Digex, which was a dot.com startup before its time. At the time, no one knew then what a cyberologist was, and I was worried about what I was going to do for a job—to make a living.

Early Dot.Com Internet Services Provider

Not knowing then that I was way ahead of my time, my career has mostly been based on genuine luck all the way. In leaving the Pentagon, where I worked as a government contractor, I had no idea what I was going to do. When Digex called, I started there as employee #50. Digex was an early U.S. internet services provider in the managed hosting business—a basement startup.

Digex were hosting servers for dot.com. Although at the time, I did not know, Digex was a start-up pioneer serving subscription pornography services. Their servers sat over top of a Chinese restaurant. Soon after, dot.com was starting to really take off for clients in data centers, banking, online, e-commerce, and new emerging businesses.

P. Fusco (✉)

D.A. Christofferson (ed.), *Women in Security*, Women in Engineering and Science, https://doi.org/10.1007/978-3-319-57795-1_10

Digex hosted ecommerce and intricate platforms for clients in the data center, including NASDAQ, Bank of America, and Novartis clinical trials. We supported critical assets for these client companies. In my role, we were also tasked with logical and physical security—one of the reasons why companies were using Digex as a hosting provider. Legislation and regulatory mandates were also starting to surface. Digex was growing very fast. My responsibilities as a CISO also included privacy and compliance, technology and building out data centers on a global level.

An IPO—Initial Public Offering—followed. Digex was sold shortly afterward for $6.1B to WorldCom, and I did very well on the sale. Golden Handcuffs was a term I became familiar with! Of 1800 employees by then, only 26 were asked to stay. People had to reapply to their jobs at WorldCom. WorldCom already had a CISO, and I did not think I would be one of those 26 asked to stay. When I was called into the office, I was thinking it was the end of the ride. Instead, I was one of the 26 asked to stay—I was floored! WorldCom paid me 4 years' salary to stay for two, presenting me with a huge financial incentive to stay.

WorldCom's primary reason for purchasing Digex was because they had so many data operations centers under the certification umbrella. I had taken the task of certifying the nine data centers for ISO 27000 (BS7799 at the time), and building out firewall operations for security. This did very well as a revenue generation entity of Digex.

From Wikipedia: Digex was as an early provider of Internet services in the United States credited with creating the "managed hosting" business, serving dial-up Internet access and web hosting farms to businesses. It was a basement Internet startup that grew to dominate and offered 3 IPOs under the same company name.

Big Pharma

My next step was moving to Merck Pharmaceuticals as CSO (Corporate Security Officer). Pharmaceuticals mostly come down to drugs and money. My role involved working with the company's drugs as the primary line of business, focusing on drug fraud, and building out a company inside of Merck. That organization was a collaborative effort of eight pharmaceutical companies. SafeBio Pharma was collaborating between Merck, Johnson & Johnson, Glaxo Smith Kline, and five other like companies.

Digital signatures were needed in the industry to support clinical trials, physicians, and other aspects of the business. Each company initially contributed $500,000 each, which was turned into $1,000,000 each, or a total $8,000,000 investment to build out SafeBio Pharma. I founded a Board of Directors for SafeBio. As part of that charter, I sat on the board, to create digital signatures for doctors and clinical scientists, which also had to be supported by the U.S. FDA—Food and Drug Administration. SafeBio hired a CEO (Corporate Executive Officer), CFO (Corporate Finance Officer) and was managed by those eight contributors. The standalone company that was managed by these eight companies still exists today, that was built in 2004. They issue digital credentials online for doctors and pharmacists. It was a very successful effort. This was a part of my day job that I was not expecting, but it was very rewarding and I really enjoyed it.

Banking and Financial

Changing industries again, my next role was with Citigroup as Executive VP (Vice President) of Information Security. A team of 400 reported into my organization. Citigroup was suffering from quite a few management summary findings from auditors on lack of security. Having also recently purchased Solomon Smith Barney, they were now the biggest bank in the world. With increasing scrutiny from regulators, Citigroup had much do in a short period of time.

Managed Security Services

Originally hired into a conglomerate made up of FishNet, SiegeWorks and TrueNorth, it soon became just FishNet. The founder of SiegeWorks was asking for help pulling these three companies together, to ascertain what service offerings would serve clients globally. My role helped build out the product service line and customer base.

Next stop: Working with Solutionary, another managed security services provider, where I did much of the same activities. The company was merging with several other organizations.

For Profit Education

The Apollo Education Group hired me as CISO to lead their security function. At the time, the sector was under regulatory scrutiny, as a U.S. public company receiving government backed student loans.

Startups

Recently, my role has included working with several startups to help successfully launch, and ultimately be acquired as part of industry mergers and acquisitions activity. It was exhausting!

It's All About the Cocoa Beans

With Digex having so many customers under their umbrella, I had to understand the large varied customer base and what was most important to them. For example, Phillip Morris's product base extended far beyond tobacco products and included Clorox and kitty litter.

As a pharma company, you might think that getting drugs to market was most important. Clinical trials were actually the most coveted. It mattered significantly what they were doing with other pharma companies to merge drug recipes to take to market, to create a more solid end state patient solution.

With finances, of course they are important; it is money that matters most.

Nestlé's primary focal point for security was Cocoa Beans, and sites that monitored the weather.

You really have to look at the nucleus of company—what gets the products on the shelf. If the cocoa beans crops were not producing, then Nestle lost money. Most important to Nestle, were the weather servers monitoring cocoa beans.

Understanding the company's business is key, and knowing the client and their line of business.

Security is not about company's architecture or their people. It is about understanding what is core to that business.

Being exposed to this early on, proved to be a differentiator. Being thrown into it, forced me to learn it from osmosis. Many people never figure that out. Learn from your mistakes and take it to the next level.

Showcasing Technology and Innovation

Diversity of Clients, Business Lines and Employers

My career has exposed me to many different business lines, in industry-leading environments:

- With my early navy years, cryptography was ingrained early. It has remained a consistent and evolutionary solution to protect transactions, data and privacy.
- Working at Digex, with so many different clients, supporting a digital economy as it breathed early life: E.G.: J. Crew started their online catalog sales through Digex, which was very dynamic and innovative at the time. No one was doing it. It began right here also, with the USPS—United States Postal Service—online purchasing of postage stamps, through their data center.
- Having broad diversity in my employment base: Pharma, manufacturing, healthcare, finance, higher education, and technology startups.
- Dealing with regulatory requirements: PCI with Visa and MasterCard, and pulling HIPAA together in Healthcare (Pharma).

- Adopting standards early on for ISO 27000, to certify systems.

When managing so many different clients, that are so diverse, on a global level, you really have to design technology that meets the needs of many. Yet it must also support that indigenous market share, whether it falls into healthcare, finance, manufacturing, or any other business line. You must design an environment that meets the needs of the many.

80/20 Infrastructure

Eighty percent of what a customer received when signing up with Digex was the same across the board. Twenty percent was different on top of that, to meet the vertical—to meet unique client needs.

We instituted that globally. We were able to maintain asset inventory—what systems they had under their roof, what applications were being used, and what transactions were taking place.

When an event occurred—a network outage, or something to do with security, we knew what we had in our data operations centers.

In building commerce servers, they did not build one off per customer. Instead, they build five versions of a commerce server build, and the client could pick from one of those five. Five versions—not 36,000 versions. It was easy to quickly scan and look for an exploit out there on a server, and of those five builds, see how many of them had the vulnerability. It made it easy to ascertain how many servers had to be secured. This really reduced the number of servers that had to be secured, vs. having to look at and secure the total.

The Golden Era of Internet Outages from Virus Attacks

Internet outages were the beginning rage: Slammer, Blaster, Code Red, Nimda—were very big at the time. They were impacting most companies negatively and creating major outages that took many man hours to address.

At Digex, they put in many hours to design what they did in their data centers at the global level. The goal was to deploy a "1-to-many" solution (vs. 1:1) with remote management in ninety operations centers: In Tokyo, Germany, Paris, London, US east and west coasts—everywhere.

Slammer was one of the largest exploits hitting companies around the world at the time.

Two and half years into their design and build-out, 7 PM on the US east coast:

At 7 PM, walking out of the building with some of my staff, we were stopped by President of the company.

He asked: "Where do you think you're going?"

I'm thinking, "My God, as if I haven't put enough hours in over the last three or four years."

Me: "Well, we're leaving, it's Friday, and we're going home."

Him: "Well everyone's getting impacted by Slammer." Me: Yes, I know. Digex never got hit."

The President looked at me and said: "That's what I'm telling you. How does it feel to be able to walk out of this building at 7 o'clock, knowing that everything you and your team have done over the past two years that Slammer never impacted any of our clients or Digex, but the rest of the world is suffering? You should pat yourself on the back."

We got to go home while the rest of the world was having issues and working through the day and night.

That was very meaningful to me; I never thought of it that way. I thought Wow! I guess we are doing well.

As a result, Digex received very big investments from Sun and Microsoft early on, which helped grow the company. That investment came in as $100 million each. It required Digex to build all the client infrastructures in the Digex data center on their systems (Sun & Microsoft). We were to use their applications when Digex was able to do so, which is why the investment was made.

Designing for Patching and Bug Fixes

At least once a week, the Microsoft platform was requiring a patch or bug fix very early on, for Tuesday or Wednesday. Security vulnerabilities were also coming in on Microsoft servers that had to be fixed.

At Digex, this was very difficult to do, with 50,000 servers. For one minute of unscheduled downtime per client, Digex paid five million dollars. We could not arbitrarily push patches, as it created downtime for the customer and had to be scheduled. It became constant and we could not keep up with it.

This meant redesigning server builds and understanding how we do this to:

1. Design this so we can take care of these bug fixes and take care of our clients
2. Do it in a regular maintenance window as opposed to all the one-offs, which just was not scaling.

Considerations in the redesign:

- Where to put the operating system and applications and all the data that resided with them
- When Microsoft pushed a patch, we needed to know which systems would or would not need the patch

We would then run a security script, and in less than an hour, have an absolute inventory of last known state of that server, to determine if it needed to be patched, and its criticality.

At the same time, many organizations suffered deeply. "Critical" patches being pushed out were not being fully tested, and they were being pushed out within a 24-hour. window. Their deployment was breaking systems, applications and technology, plus transactions and everything on the transport level.

We did not suffer that at Digex. Our service level agreements (SLAs) did not allow this much disruption. Changing our system design meant working very closely with engineering and operations teams to create a solution to meet our client SLAs.

Microsoft came to us and asked how we were doing patching at Digex and not suffering like their other clients. Microsoft was making an assumption, that clients were keeping their operating systems and applications just as delivered—with partitioned drives using standard Microsoft naming conventions. Digex did not make that assumption. That assumption was: Patch, push, just do it, and it will fix what we put on the "C" drive. But "C" was not necessarily the name of the drive the customer had.

Digex was not doing it that way. At Digex, we looked at the absolute data/time stamp for last known state for any patch, for any drive in the system, and we patched it that way, not by drive name.

If it is not taking, then you look at variables for how people CAN change their systems. People can configure their drives however they want, to meet their needs. Microsoft assumed that the drives were kept the same way they were delivered.

As a result, Microsoft re-engineered their whole patching system.

Fast and Furious

Those 7 years created the foundation for so much that happened in our industry from a cyber standpoint:

- The beginning and onset of dotcoms, the rise and fall of dotcoms.
- Year 2000 (Y2K and the data turnover in systems)

It set the stage for enormous learning in such a fast and furious time frame. It helped propel me. Although I am always looking for another Digex, I have not yet found it. For me, having it early on gave me an awesome career. Not being able to find it again has hurt me personally. I look for that vibe and have not yet found it again.

Early Influences and Future Success

Digex profited by hosting companies and their applications. They hired some of the best talent, who were very forward thinking. They built cloud way before its time.

At Digex, we also designed data operations centers very differently than most were designed. Called farms, inside a data center, there were nine globally, with probably ten different farms. Huge data centers resided within those farms. When designed, they put things they would need in a data operations center, like maintenance, outside of the farm. E.G., for the air handler systems that supported the data center: Maintenance staff did not come into the server area. The air conditioning systems were maintained from the outside.

In applying methods to reduce risk and oversight, a lot of influence came from my government background:

- 1-to-many builds
- Differentiation between black and red networks (CipperNet and NipperNet)
- Being a cryptologist
- Working with so many diverse people in the Military

Working Together to Reach the End State

In the Military, I would come home on a given day, and tell my husband, that I'm going for 90 days, and I'm being deployed without knowing where I'm going. You have to be able to assume things quickly and address it with a calm cool mind.

You worked with people you never knew before. You may not like them, but you have to learn to work with them, because the military never fires anybody, and lives may depend on it.

This was instituted into my work world. Working with engineers, in UNIX, applications, and developers, they may be driven differently that what drives you. You have to learn to work with them. Falling back on my Military background helped quite a bit in that arena.

We also had operations folks sit side by side with engineers, and the engineering staff sit side by side with developers. You take 2–3 days, and then transition them out. Whatever you are designing and engineering, it then goes into operations, and these same teams need to know what the impact is on operations.

This gave the Operations teams an understanding of why engineering was designing what they were, or what architecture was being created with an operations element. At Digex, we brought together these teams to collaborate together on their work.

This collaboration helped everyone understand why we are doing what we are doing. Otherwise, people do not understand and it does not put them on the same mission. It is not necessary to agree, but just to understand the end state, beyond the purpose of their individual job.

Taking this approach helps when we lack communication within our organizations or with our clients.

Evolution of Security and the CISO Role

Many people who have been in security as long as us, 30 years or so, have often come from a Military background.

The CISO role and security in general, is a very new profession. Looking back, the CISO role came into existence around 1999 or 2000. It was not well known or publicized. The CIO—Corporate Information Officer—role has existed a long time.

Those of us in security for the last 20–30 years are truly the genesis of the CISO role. We brought it to what it is, and where we are. Nobody knew it would be what it is today. The CISO role has evolved significantly. We have weeded out what it was, what it should not be, and what it needs to be.

It should not report to the CIO, but needs to be more diverse. It is not about just tracking hackers. It must also involve compliance, risk and governance oversight. Many CISOs also have physical security within their responsibility.

Not everybody has been it that long. But the longer term security professionals bring a lot to the table. They are not just average fly-by-night security people—they have been in the field a long time—since the beginning. These people are not police officers or FBI—US Federal Bureau of Investigations staff. Those roles do not bring together what a CISO or cyberologist do.

Going through 9–11, gave me a different experience that I never expected. United Airlines was a major client at the time. When dealing with terrorists and disasters impacting technology, you have a completely different undertaking. It showed how dependent we were on technology when that attack came.

United Airlines had to have unprecedented never-before-asked, immediate, access to systems and servers when no one else was online, and everybody else was knocked off. That was unheard of at the time. Yet at Digex, we were able to do it. Because of their 1-to-many build, they could build a server in 10 min. Using dynamic, forward moving, and standard builds, they were able to meet this need.

My experience in the industry really helped me. Not a whole lot of us have been in the industry that long.

Security by Definition

You hear multiple words for security and cybersecurity seems to be word of today. Is there a difference in cybersecurity, security assurance, information security, or other words to describe our field? There is no uniqueness in the term cybersecurity. It usually becomes whatever Gardner comes out with and calls it. For security across the board, the most accurate label is information security.

When people talk about Privacy, you cannot have privacy without security and you cannot have security without privacy. Security is: Governance, risk, compliance, privacy, oversight, business risk management, litigation, and forensics. It is interchangeable and all of the above. It is about information. When people want to get into systems, when we have risks or threats, or attacks, it is to get to the information.

We call it cyber because we are living in a virtual world and cyber is more eclectic to the twenty-first century. For instance, I am a cryptologist by trade, and it is about cryptography.

Getting into Security Today

Many opportunities exist. When someone says to me and says: "I really want to get into security", I answer: "Great", regardless of what field they are in now. "What part of security do you want to get involved in?"

This is where people can go astray. They like the concept of black hat white hat, and the spy piece of it. If they do not understand what makes them tick, they will not be happy in the security arena.

So I ask them: "What part of security do you want to be in?" They get that dog-cocks-it's -head look, like "What?"

Do you like networking? Applications? Policy? Compliance? Legal? Are you a developer? Do you have a different mission? Do you want to write code? They must truly understand what makes them tick.

If you want to get into security, it is so much more than you might see at the surface. As a security individual, you have to understand in the middle of a breach, or a morphing Trojan, or whatever it is: Where it is coming from? Is it the network, the application, an individual, a server, a botnet, or is it something we have not seen before?

You have to understand technology across the board. You need to understand application. If you do not understand, then you cannot ascertain what the criticality is.

Security is not indigenous to one area. When I look at people in IT—Information Technology, you have the server administrator, the network administrator, the firewall administrator, and each is indigenous to those specific areas. With security, you have to quickly ascertain, is it the network, firewall, application, user or is it an outage? You have to know it all.

By default, security people are better-rounded when it comes to IT operations, because you have to be. You cannot do just one thing. You have to know what it is.

So when someone is asking me, if they want to get into security, I always try and see what makes them tick.

Transitions to Security

For careers in transition, or individuals coming to me who want to transition into security, let's start with someone in the healthcare industry.

Nurses for example, who want to get into technology and security, have an essential background in the security arena, when it comes to healthcare, HIPAA, and electronic records. I ask them if they like compliance and policy as it relates to healthcare. If they say yes, I tell them that this is a nice niche for them.

You can come into the security side, under Governance, Risk, Compliance, GLBA (Gramm-Leach-Bliley Act), patriot, healthcare, electronic transactions, now going into wearable devices, and insulin pumps or pacemakers that are remotely managed.

You look at what people are doing, where the world is, and what we use technology for, and then I can say:

> "Almost anybody that wants to do security, regardless of what your field is today, can do it. You could be a mechanic, dentist, nurse, teacher, or bookkeeper. If you want to do security, there is definitely a niche for you, using the experience that you have, along with your background. People often just do not what is out there."

What is out there right now? Cloud computing is huge, business intelligence analytics are back in play big time when it comes to security, and business intelligence from those analytics.

To help people who want to come into security, we have to take the time to understand what they have done, or if they are just out of college, and what make them tick. Before we tell them to be a security network engineer, or a security applications developer, we have to understand them, so that they can do it for the long term. That is where we are falling down.

To help others succeed, I always give back. Because I think one of the essential parts is: Cybersecurity is my profession and wherever I'm working at that time, that is my day job. But my profession is cybersecurity.

Keeping that in mind, you always want to understand what is happening outside of just your corporation. So if you're working in healthcare, and you are only focused on healthcare, you are not going to know how finance is managing security or manufacturing, or e-commerce.

Paying It Forward

Paying it forward and being involved in conglomerates, and associations, and in collaborative efforts have been my mainstays. That is not because of my job, it is on my own, because I want to be involved in the profession.

Along the way I created other consortiums, as the industry started to blend in with the rest of world.

Many years ago, I became a member of ISSA (Information Systems Security Association) and started the CISO Executive Forum. As a founding member, I also helped Jim Reavis spearhead the Cloud Security Alliance, and I still belong to the CSA today. In addition to a full-time day job, I believe in paying it forward in the industry. This helps me understand what is going on globally in cybersecurity, vs. a singular internal viewpoint. These have helped propel me quite a bit along the way.

The #1 key for anybody, is to be involved in the profession. I really look for that in people I hire for teams. "What is it that you do outside, to support your profession, that is not just your day job?".

In talking to people: "Do you want this to be your profession, or is just a day job for you?" Differentiate the two.

If this is your day job, it is not going to last very long for you. You are probably going to get burned out, you are not going to be happy, and you are probably not to get the big picture. But if you tell me you want information security to be your profession, I'm all in, to help you get to where you need to be.

Who Helped Me?

Quite a few folks helped me.

- Admiral McConnell, the Director of NSA, who I worked directly under him when I was a young sailor.
- Mike Scagnelli, Captain in the US Navy, Cryptologic officer, helped me along the way.
- Mark Scholl a CEO at Digex trusted me. In such a quick moving environment, trust was important during incidents, and for buy-in for budget, business cases, etc. He trusted me to have it fully baked.

My experience has been good, because I had really good people to believe in, trust and support me along the way. If I had not asked for what I needed, I would not have achieved the career I have today. In return, I put my trust back into them, and upheld my commitments. Getting my start in Operations helped me to understand technology.

Women were not part of the path—I did not have a woman who helped me along the way. There were not a lot of females in the industry, and our numbers have not changed much over the years. One exception was Grace Hopper.

Amazing Grace

One of the highlights of my career was meeting Grace Hopper. She was called Amazing Grace and is since deceased. She was the first female Admiral that the Military ever had. It was stunning to meet her. Grace made a very profound statement: "People wake up! Soon data will be an asset, and you need to start tagging your data as an asset." And she was right, back in 1992. I was in the Military at the time, and Grace Hopper was retired.

Personal Crossroads

At the time, I did not realize that I was building a foundation from everything I did. I only aligned the dots as I got older. You only realize it when you talk about it—like now in this book. Being in the industry so long, and having sat as CISO multiple times, I have hit many bumps and cross-roads.

You come to a crossroads of working with organizations or entities that are meaningful and purposeful in your life, and not just a paycheck. The company must be ethical and moral.

Many of us struggle to find that entity that is out there, that needs and wants security, to leverage it in the right way, and not just cross the T's and dot the I's, and check boxes. That is not who or what we are in security, and it is not who I want to be. It is hard to fit in the industry because of it. Experience is great on one hand, but it creates a personal struggle.

Look back and look forward. Identity the entry points today and going forward, which you can leverage and take advantage of the opportunities. Then go for it.

Nobody Owns the Moon

The next evolution for me would have to be in identity management and identity credentials. We cannot tell a 5-year old from a 50-year old on the network. That is number one, where my passion lies, in creating unique identifiers. This involves not just a social security, which is unique only to the United States, but for everybody on the Internet globally.

At my next level I see myself pulling together disparate solutions. We need to rethink what we are using, how we are using, and what we are doing. It means that we take a step back and look at who is running the internet. No body owns the moon, and if you were a rocket, you could fly there anytime you wanted, without anybody's permission. Nobody owns the Internet.

It will be a long time before that happens. A majority of individuals and the users of the Internet are common consumers: You, me, my mother, brother, aunt or uncle, or anybody. Looking at and understanding what the individual does on the Internet—why they are doing what they are doing, and why they need that access—sets precedence for the next evolution of security in business.

When we look at something like the Apple iPhone, it became a critical asset, because consumers, common lay people, brought them into the office and wanted to use them. The iPhone got adopted by teenagers, preteens, and grandparents, parents, and soon everyone was using an iPhone. It was not driven by big business. It was driven by the common consumer.

Looking at what we do as human beings on a daily basis, will drive the next evolution of what needs to come for security. This means we can get ahead of the game, and see what the critical elements will be, as we move along.

My Next Mission

That perspective is driving what I am doing today. My mission right now is to look at many of the solutions we have, such as firewalls, and ask, "Do we need them all?" We are evolving to a place where we no longer need all these solutions to support security.

If we understand why a human being is doing what they are doing on the Internet, and why someone would want to exploit, or gain access to, or garnish information, then we can establish security parameters. It truly goes back to the individual.

Number one is identifying who, what, and where, and then the why and how of it all: Who did it? Whose is it? Should it have been touched? … and so on. It takes us right back to the individual as a unique identifier.

My next evolution is a unique start-up company to pull that together. It also will include work on breach notification. Solutions today fall all over the board. We need to come to a common denominator, which would be more like a 90/10. That means that 90% meet this requirement, and 10% is indigenous to the different elements as it relates to breach notification.

That is my mission today.

Chapter 11
Organizational Intelligence: Cybersecurity as a Performance Optimizer

Rhonda Farrell

Recent cybersecurity innovations in both technology and standards, set a new tone for organizations, allowing for cyber-intelligence to be capitalized on by focusing on continuous collection, analysis, policy enforcement, and remediation at multiple enterprise levels. Vast improvements in governance, risk, and compliance management follow.

Focusing on these three areas allows for enhanced collaboration with strategy, change management, and quality efforts, allowing for heightened performance and personnel optimization.

Organizational Intelligence

What if the operating model were changed to use cybersecurity-related capabilities as the model for organizational performance optimization efforts?

Is it possible that leaders could drastically transform enterprise practices, by better enabling cyber initiatives to lead the way by focusing on integrating capabilities, tools, and data?

Would organizations be willing to enhance their governance, risk, and compliance efforts by incorporating model elements from the cyber realm?

More importantly will cyber leaders and executives be willing to step out of their comfort zone and examine the rapidly changing worlds of strategy, change management, and organizational excellence for ideas, practices, and quantitative method-

R. Farrell (✉)

D.A. Christofferson (ed.), *Women in Security*, Women in Engineering and Science, https://doi.org/10.1007/978-3-319-57795-1_11

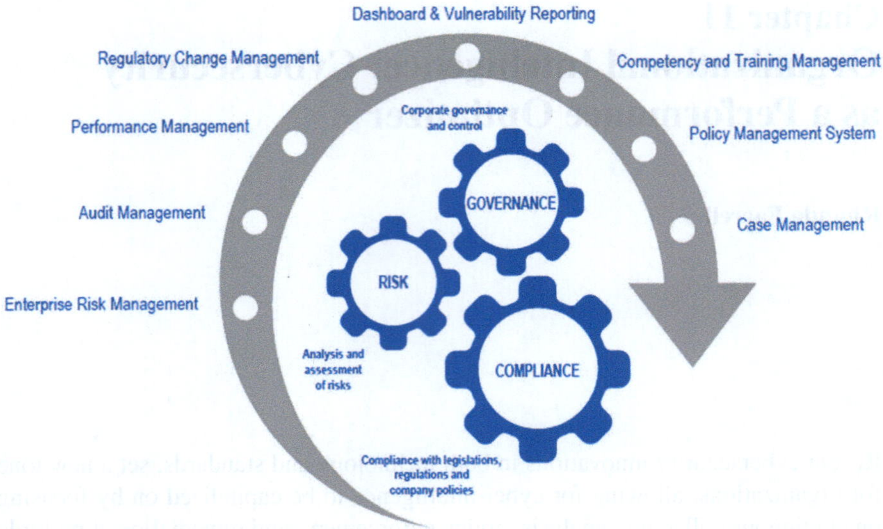

Fig. 11.1 Typical GRC suite of services (http://www.360factors.com/information-security-risk-management-software/vulnerability-risk-management/)

ologies that will raise the bar significantly in the areas of engagement, ethics, value, and sustainability?

Figure 11.1, below shows a typical Governance, Risk, and Compliance (GRC) services suite which focuses heavily on standard organizational elements which allow for organizations to identity, manage, and remediate risks across the enterprise.

The components listed below offer a solid baseline to work from and offer plenty of maturation opportunities, as the focus turns to WHY, HOW, and to WHAT EFFECT initiative implementation will have on the organization, products, services, stakeholders, and most importantly, the workforce.

Getting leaders and practitioners to focus on the WHY behind the action tends to be the game changer—as this allows for the core components relating to the organization's people, process, and technologies to be identified. Once these basic elements have been thought through and enumerated, further strategic alignment activities can occur to mission, vision, goals, objectives, and key performance indicators.

Taking a methodical examination of the enterprise interconnect points, oftentimes paves the way for enablement of the widest possible re-use to maximize value across the entire organization.

These initial discovery, planning, and strategy undertakings often leads to the publishing of initial outputs, allowing for the larger stakeholder base to emulate the "WHY" exercise at the local levels, thereby unlocking innovation and performance opportunities and enhancing organizational business intelligence capabilities, per Fig. 11.2.

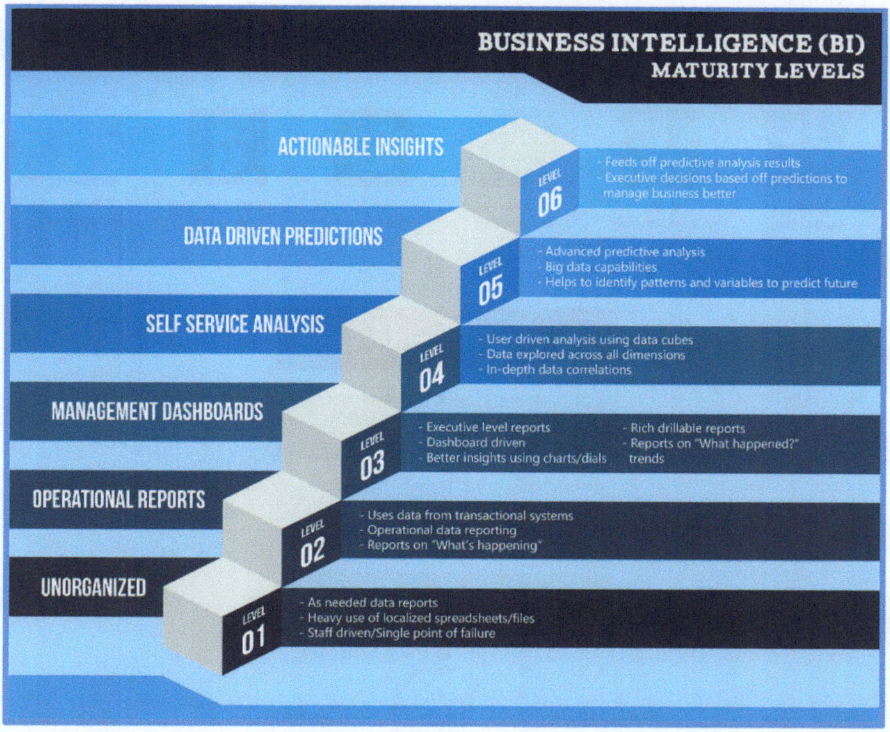

Fig. 11.2 Business intelligence maturity model (http://www.ventera.com/news/insight/white-paper-sensible-approach-successful-business-intelligence-biimplementation)

The focus on maturing cybersecurity related GRC initiatives oftentimes provides for rich data deposits that can be mined for product, service, infrastructure, and business capability creation purposes. Additionally, it also helps identify workforce densities and geographic disbursement patterns, thus opening the door to further knowledge management, collaboration, and information sharing efficiencies.

Examining cyber generated data through the business intelligence lens and continually asking how the data and information can inform, focuses not only on hardening the WHY, but also addresses the HOW behind making the initial operating leap to using a constantly questioning, analyzing, and innovating, data driven approach.

As Fig. 11.2 shows, incorporating a business intelligence model informed by cyber generated information can lead to a much wider perspective on value creation across the enterprise. One of the results could be a more robust GRC services suite, augmented to include these new capabilities, much like Fig. 11.3 below.

Primarily the reader will notice that with the addition of the business intelligence element, the product and service capabilities can then also drive towards answering the TO WHAT EFFECT inquiry that must be done to drive further efficiency and effectiveness gains (see Transformation related activities).

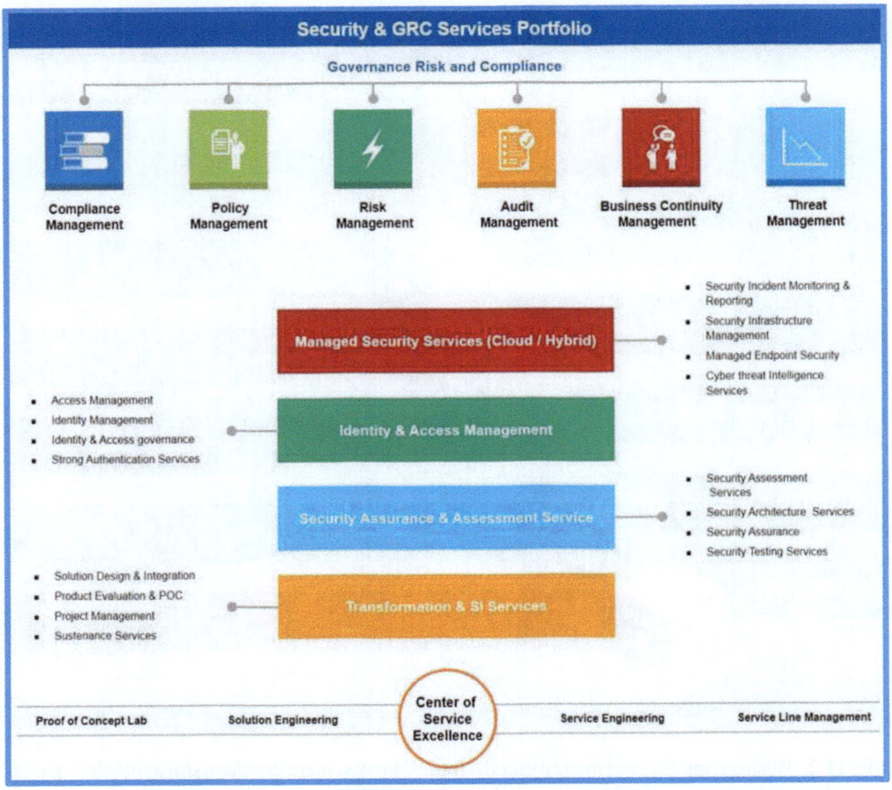

Fig. 11.3 Enhanced GRC suite (https://www.hcltech.com/it-infrastructure-management/information-security-services)

Even though the rudimentary basics of WHY, WHAT, and TO WHAT EFFECT are being consistently addressed by using quantitative analysis on risks, compliance assessments, and corporate governance findings, several key programmatic components may still be lacking or immature, disallowing or impacting organizational leaders in their quest to take their programs to the next level.

Foremost among these highly optimizing program elements are performance management, solid leadership practices, workforce engagement, utilization of enhanced knowledge management systems, and most important, high involvement customer/client management.

Figure 11.4 outlines the core process components as extracted from the Baldrige Excellence Framework (2015–2016). For those non-familiars, this body of knowledge focuses on a "systems approach to improving your organization's performance" as well as undertaking actions which "empower your organization to reach its goals, improve results, and become more competitive" (pp. title, i).

Fig. 11.4 Baldrige—organizational excellence—key process perspective (http://www.baldrige21.com/Baldrige%20Model.html)

Not only does this body of knowledge help enterprises solidify their WHY, it offers a flexible methodology for ascertaining a set of HOWs approach, as well as incorporates a heavy Performance Management focus, which helps organizations harden their TO WHAT EFFECT inquiry and projections.

In addition to expounding upon the core Baldrige processes as outlined in Fig. 11.4, the National Institute of Standards and Technology (NIST) recently released the Baldrige Cybersecurity Excellence Builder, which informs key cybersecurity leadersVhip and practitioners on key organizational excellence elements they can put in place to harden their overall cybersecurity programs, while at the same time strengthening the bonds between the realms of cybersecurity and quality overall.

Figure 11.5 depicts the organization cyber-related roles associated with expected use of this new guidance, as well as benefit enumeration which can be achieved organizationally, thus bolstering the case of the addition of heightened business intelligence capabilities to inform enterprise decision-making on a wider scale.

Once the core Baldrige building blocks have been put into place the harder work of ensuring strategic alignment within the organization must then be used to tear down silos of inefficiency and pave the way for transparent communication and collaboration efforts.

Who in an organization should use the *Baldrige Cybersecurity Excellence Builder*?

The *Baldrige Cybersecurity Excellence Builder* is intended for use by the leaders and managers in your organization who are concerned with and responsible for mission-driven, cybersecurity-related policy and operations. These leaders and managers may include senior leaders, chief security officers, and chief information officers, among others.

Role/Function	Benefit of/Reason for Using the *Baldrige Cybersecurity Excellence Builder*
Board and Executive Management	• Understand how internal and external cybersecurity should support organizational (business) objectives, including support for customers • Understand current and planned workforce engagement processes and their success • Understand opportunities to improve cybersecurity in alignment with organizational objectives • Understand the potential exposure of the organization's assets to various risks • Align cybersecurity policy and practices with the organization's mission, vision, and values
Chief Information Officer (CIO)	• Understand how cybersecurity affects organizational information management practices and culture • Improve communication and engagement with organizational leaders and the cybersecurity workforce • Understand how cybersecurity affects the organization's culture and environment
Chief Information Security Officer (CISO)	• Support the organization's commitment to legal and ethical behavior • Create and apply cybersecurity policy and practices to support the organization's mission, vision, and values • Respond to rapid or unexpected organizational or external changes • Support continuous improvement through periodic use of the self-assessment tool • Support organizational understanding of compliance with various contractual and/or regulatory requirements • Understand the effectiveness of workforce communication, learning, and engagement, as well as operational considerations for cybersecurity.
IT Process Management	• Improve understanding of business requirements and mission objectives and their priorities • Determine the effectiveness of IT processes and potential improvements • Understand how aspects of cybersecurity are integrated with organizational change management processes
Risk Management	• Discern the impact of cybersecurity on internal/external customers, partners, and workforce • Improve understanding of how workforce engagement in cybersecurity and communication to the workforce about cybersecurity impact the organization's overall risk posture
Legal/Compliance Roles	• Understand legal/ethical behavior on the part of the workforce, as well as the overall cultural environment • Understand how the organization applies cybersecurity-related policies and operations to ensure responsible governance, including legal, regulatory, and community concerns
Employees (Workforce)	• Understand leaders' expectations • Be better prepared for changes in cybersecurity capability and capacity needs • Benefit from a workplace culture and environment characterized by open communication, high performance, and engagement in cybersecurity matters • Learn to fulfill their cybersecurity roles and responsibilities

Fig. 11.5 Baldrige—organizational excellence—key process perspective (https://www.nist.gov/sites/default/files/documents/2016/09/15/baldrige-cybersecurity-excellence-builder-draft-09.2016.pdf)

Often-times these initiatives focus on enhancing the mission and vision of the organization, enterprise, or business unit. It may also entail defining new goals and objectives which focus on heightened mission achievement, flowing from the enhanced strategy, which in turn inform new or updated policies.

Innovation and technology capabilities are then oftentimes acquired or developed to provide enhanced analytics surrounding programmatic and technical risk to quantify likelihood or impact with increasing rigor. Additionally, cyber hardening initiatives oftentimes focus on enhancing:

(a) workforce capabilities to meet mission need;

Fig. 11.6 Organizational excellence building blocks for Agile organizations (https://www.linkedin.com/pulse/organizational-excellence-building-agile-organization-gary-harpst)

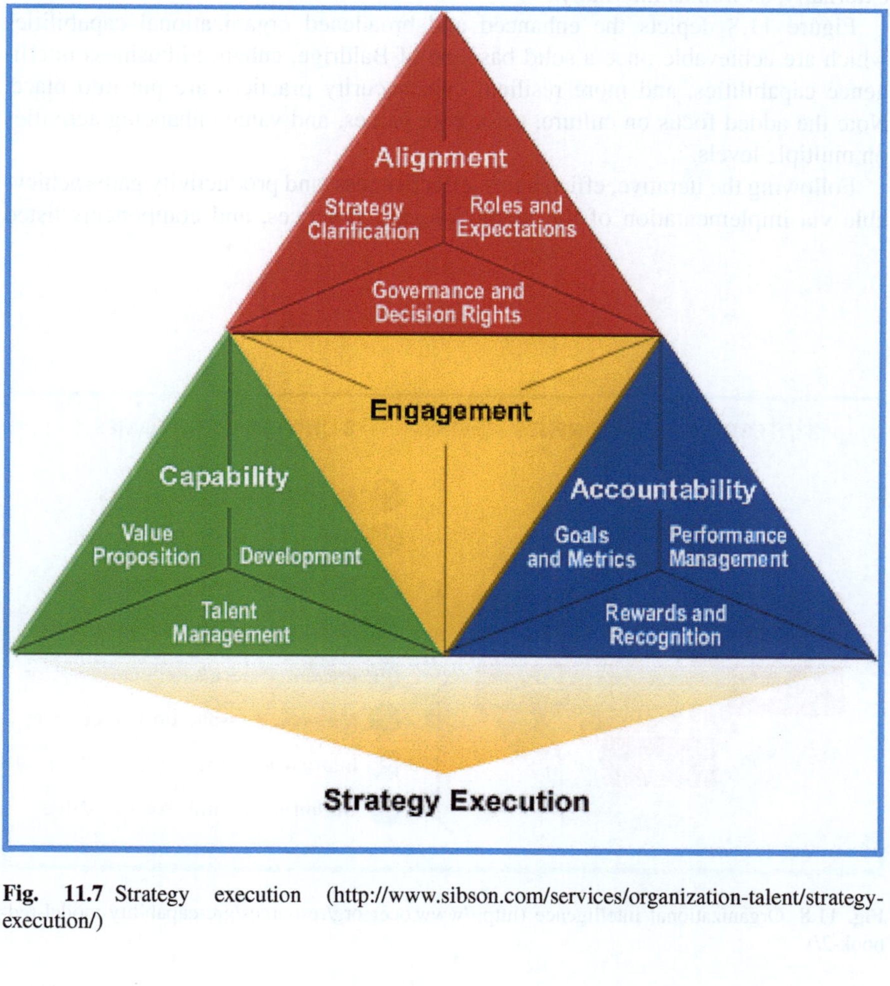

Fig. 11.7 Strategy execution (http://www.sibson.com/services/organization-talent/strategy-execution/)

(b) wide-spread and in-depth process improvement; or
(c) business process re-engineering to better capitalize on the plethora of findings generated by the business intelligence algorithms. Figure 11.6 outlines the organizational excellence elements necessary to enable programmatic success.

Figure 11.7 shows an example of a more robust strategy execution model, depicting the strong ties between heightened governance practices, high levels of workforce and leadership engagement, value-laden capability development, paired with stringent performance management practices, and generous rewards and recognition when high levels of mission achievement are the ultimate result.

Once the core Baldrige components have been steeped from both an organizational optimization and cybersecurity perspective longer term continuous improvement program elements can then be put into place. This entails examining the organization holistically, at a far deeper level, focusing on triaging trouble-spots, innovating through difficulties, and providing enriched value-add capabilities, both externally, as well as internally.

Figure 11.8 depicts the enhanced and broadened organizational capabilities which are achievable once a solid baseline of Baldrige, enhanced business intelligence capabilities, and more resilient cybersecurity practices are put into place. Note the added focus on culture, trust, core values, and value enhancing activities on multiple levels.

Following the iterative, efficiencies, effectiveness, and productivity gains achievable via implementation of the methodologies, practices, and components listed

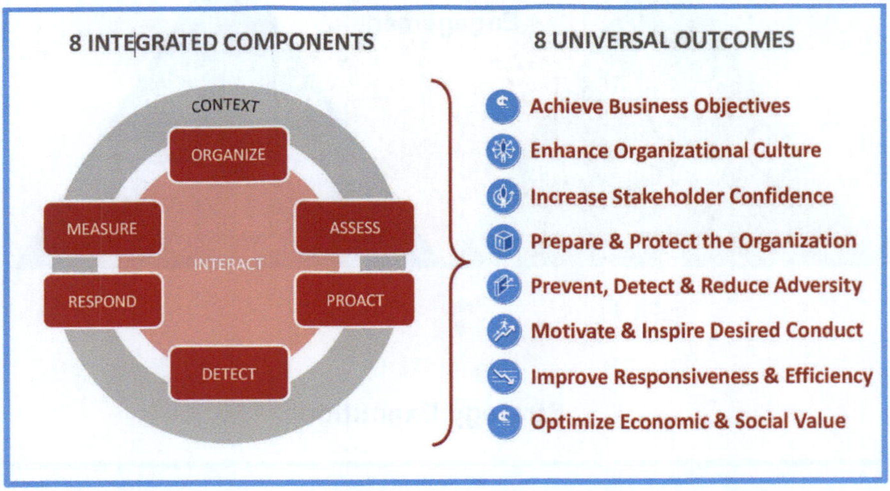

Fig. 11.8 Organizational intelligence (http://www.oceg.org/resources/grc-capability-model-red-book-2/)

Fig. 11.9 The key model elements (http://www.bendelta.com/news/strategy/effective-leadership-important-for-strategy-execution/)

above, Fig. 11.9 shows the highly recognizable key elements associated with driving continual success for your organization.

Never fear, as once the enterprise has achieved this interim plateau called ‘success’, where everything makes sense, at least for some modicum amount of time, it’s almost certain that industry will innovate beyond, requiring the organization (and you) to rethink your WHY once again.

Never doubt the process, for as the organization keeps shooting for the moon, they’ll reach the STARS a million times over, raising morale, building momentum, and creating sustainable motivation to continue to the next level of maturity and performance optimization.

Here’s to much Baldrige inspired success for you and your organization!

Part III
Foundation for Newcomers

Chapter 12
The Pursuit of Cybersecurity Career

Juanita Agard

Introduction

I am profoundly grateful for the opportunity to use this platform to share personal experiences—all of which have in their own ways culminated in a career in cybersecurity. To be honest, my journey has, at times, been riddled with challenges and upheavals. However, those experiences have tought me to be steadfast and relentless in my drive to finish what I start.

Full-Time Parent and Full-Time Consultant

Being present and attentive to the dynamic and unique demands of being both a full-time mother, a full-time IT Management Consultant, while continuing my eduation in cybersecurity, cannot be underestimated.

It has indeed been a beautiful struggle to show up constantly and consistently as the best version of myself in these various roles. My uncompromising love for family, coupled with a high degree of self-discipline and compelling drive to achieve my cyber goals inspired me to stay the course.

There is indeed a humbling joy when one succeeds in their mission, and the dream becomes reality.

J. Agard (✉)

D.A. Christofferson (ed.), *Women in Security*, Women in Engineering and Science, https://doi.org/10.1007/978-3-319-57795-1_12

Part 1: Trust Your Instincts and Embrace Your Natural Tendencies

From a very young age, I established myself as an observer of people and their behaviors. Though I did not think of it in quite those terms at the time, in retrospect, my early musings in investigatory work were palpable. As the only child, I was shy and quiet. I had a lot of time on my hands while being left to my own devices. Me, being so quiet, people often forgot I was in the room, which allowed me a covert, and happenstance vantage point to hear many mature conversations, observe behaviors and patterns, and collect information about everything in my environment.

I also enjoyed investigating—I was curious about everything; who ate the last piece of pie, who spilled that water, who slammed the door, who was it that my mother was talking to on the phone, who was that person putting envelopes in the mail box, how did that big spider crawl through that tiny space?

My keen interest in observation and my reconnaissance tendencies served vital in the development of my personality, social and situational awareness, and eagerness to explore while being undetected. Throughout the years, I would find myself collecting information to predict tendencies and outcomes—while thoroughly entertaining myself. I thrived in the pursuit of details, the patterns that were gleaned, and their effects on outcomes.

While being passively developed throughout my childhood in various scenarios, overtime the skills I had acquired and practiced in sport began to play a larger role in my character development too.

In college, it was evident that engineering and math helped me apply rigor to patterns and understand cause and effect relationships. Not to mention, the constant observation and predictive outcomes I'd engage in really improved my memorization.

These situations profoundly influenced the fine-tuning of a real skill-set; as I became more aware of my interest in deliberately using those skills, so too did I find that I had incisive social and spatial awareness, and my once passive desire to explore while being undetected had turned into a real talent for being unassuming and inconspicuous.

It wasn't until I reached my teenage years, that I realized not only was I watching people, but people were also watching me, and drawing their own conclusions. Though this realization was slightly derailing at first, it was ultimately advantageous to me as it forced me to hone in on my spatial awareness and tap into my investigative instincts. My imagination began to grow.

I developed my theories about corruption, terrorism, drugs, human trafficking, and organized crime. Often, I found myself engaged in research, trying to understand outcomes. I began devising strategies to predict outcomes. The technology was enabling me to gather information and facts with speed and draw conclusions based on evidence. Again over time, I found it thoroughly entertaining to use my talents to collect information and to protect myself from others that didn't have my best interest at heart. I thrived in the pursuit of details that contributed to outcomes. I was able to predict scenarios, in advance, to prevent harm devised by others.

Part 2: Technology Major with a Bachelor's of Science in System and Computer Science

In my early 20s, while assessing my strengths, abilities, and interests in order to settle on a subject area of study in college, I ultimately decided to major in technology. My inquisitive mind and proficiency in math and science helped me realize that majoring in computer science would complement my talents wonderfully.

I applied to Howard University's (HU) System Computer Science program and was accepted. Studying computer science seemed perfect for me. The major had just the right amount of math and opportunity to create, even if it was in computer code. At HU, I became fascinated with cybersecurity.

I was drawn to 'hacktivism' which is the act of hacking, or breaking into a computer system, for a politically or socially motivated purpose. I felt a strong connection with the principles of hacktivism—I found my passion. It reminded me of the reconnaissance work I had done early on in life.

I continue to develop my notions about everything. I would use the internet to self-educate. I realize the internet created a boundless opportunity to have knowledge at the tip of my fingers, discovering ways to benefit from using the internet enable me to advance in learning the nature of cyberspace and cybersecurity.

I graduated from HU with a Bachelor's of Science in System and Computer Science. I was grateful to complete my undergraduate degree and to mastered the fundamentals of technology and computer programming. However, my burgeoning interest in the concept and principals of hacktivism though, had not quite been satiated. At this point, my main priority was preparing myself to begin a career in Information Technology (IT).

Starting Career as an IT Specialist at IBM

Soon after graduating from Howard, I began my career as an IT Specialist at IBM. My position enabled me to work with different clients who had varying IT interests and issues. I was part of a team, and I helped clients achieve their IT goals and mitigate their IT issues.

The first client I supported was the National Institute of Health (NIH). On this particular project, my role and responsibilities on this particular project involved gathering pivotal software requirements and testing the new NIH electronic grant application.

While working on the NIH project, I subconsciously gathered requirements from a security and prevention stand point to 'break test' the electronic grant application. This testing approach intentionally interrupted the grant application process, which identified opportunities for process improvement. This was the surest way to find out what defects or glitches needed to be fixed in the code to restrict invalid or unacceptable inputs. I would soon discover this was a huge win! I was invited to work with the electronic application developers to modify the code and fix the flaws.

Few Women and Little Diversity

Soon, I became painstakingly aware of the small percentage of women on the NIH project performing in IT engineer roles. Specifically, the female to male ratio was one woman to five men on a project with over 50 team members. Being part of the team and witnessing such disproportion presented a set of unique challenges.

As a woman of color, this observation came with its set of qualms, and would often leave me feeling anxious and underrepresented. Here I was, a young woman, working on all-male teams, feeling like I was not always heard or even respected. Despite the anxiety I felt, I knew I couldn't let this defeat me. So I decided to channel this anxiety into encouragement to continue to gain the IT security work experience that I desired for myself.

Part 2: Strengthen Your Personal Vision

As I continued gaining IT work experience, my initial instinct observation of the underrepresentation of women in the IT and cybersecurity world continued to be affirmed.

At the same time, I was building confidence in my abilities as an IT engineer. This observation became an obsession of sorts of mine, one that I was able to draw inspiration from, as it fueled my ambitions to advance as a woman, as a minority, as a demi-pioneer. I felt indebted to myself to overcome the intimidation that plagued me as I came when coming to terms with the hard facts; the overwhelming ratio of successful men to women in this field– appeared to be significant. Being a natural competitor, this observation inspired me to become the best I could be in this field.

A Pursuit of Cybersecurity

At this point, I was confident that a pursuit in the cybersecurity field would present great career opportunities for me. I created a vision for myself to pursue cybersecurity educational courses and training to expand my cybersecurity knowledge. I envisioned uniting my IT engineering experience with technology law to become a cybersecurity subject matter expert.

Transitioning to a Software Requirement Engineer Position with Booz Allen Hamilton

Eventually, I landed a Software Requirement Engineer position with Booz Allen Hamilton (BAH), a company with a reputation for investing in their employees. Then, I learned of BAH's partnership with University of Maryland University College (UMUC)'s associated with the Cyber Graduate program purposed to build a competitive, forward-thinking cyber workforce for BAH. I immediately enrolled in the Cybersecurity Technology Graduate program at UMUC.

Cybersecurity Technology Program at UMUC

The blessings kept coming. I was thrilled to discover the graduate program course credits could be applied to the University's full Master of Science in Cybersecurity. This was my opportunity to further my knowledge and credentials in the manifestation of my educational and career goals.

Part 3: The Benefits of Sticking to the Plan

Security Certifications

In addition to the UMUC Cyberprogram curriculum, I completed the CompTIA Security +, and EC-Council Certified Ethical Hacker (CEH) boot camp training to obtain my cyber certifications. I took advantage of the time when I was transitioning from an old project to a new project to enroll in BAH's internal training opportunities. In 2015, while balancing motherhood and project work, I finally obtained my Cybersecurity Technology Graduate Certificate.

I also took advantage of various cyber workshop opportunities available in the Washington Metropolitan area, which included computer programming, penetration testing, and network engineering.

I joined the Women's Society of Cyberjutsu and I benefited from the IT security group study sessions and study material. This cyber community provided me with the support and assistance I needed to pass the Security + and CEH exams. I attended the 2016 B'More Rails Workshop for Women to refresh my programming skills. Furthermore, I plan to complete the Cybersecurity Technology master program in spring of 2017, and then proceed to study for the Certified Information Systems Security Professional (CISSP) certification.

It is indeed an ever-evolving dance of grace and diligence in advancing through my career while mothering two young boys at home, though these challenges encompass at once my beacon and my passion.

Future Plans for a Law Degree in Technology

My future plans include pursuing a law degree to practice technology law. I intend to continue branding myself in the IT security industry, and eventually building my cybersecurity empire; starting with a part-time network engineering contract and expanding from there.

Conclusion

As I grow and learn more about myself and cybersecurity, I feel confident in my choice to go against the grain and to persist through the discomfort that can exist when feeling out-numbered. I hope my experience motivates you to pursue your interests, to embrace a lifestyle of continuous learning, and ultimately to realize your dreams through practical work and incessant dreaming.

Remember to always **aim high**.

Chapter 13
Making it in the National Security Field as a Millennial Minority

Lacey Chong

Being a Minority in the US Defense Sector

As a young Chinese-American woman, I stand out in the stuffy meeting rooms of the DC national security industry. In meetings I am often the only woman of color and millennial—I am often asked if I am an Intern or whether this is my first job. I've been working in national security since 2003 so while this question is flattering and amusing, it demonstrates how few other millennial minority women decide to enter and remain in the national security field.

It can be intimidating to enter a room full of men who are older, Caucasian, often with prior military experience; it is a struggle to find mentors or senior managers who look like me. If you look at the leadership organization chart of any U.S. government agency—you will see a stunning lack of diversity. My background has allowed for some unique insights to succeeding in a challenging field. It is essential that other millennials and minority women join this industry and have a voice in the future direction of U.S. national security.

Academic Background

In 2003 after graduating with a B.A. in Asian Studies, I began my career working for U.S. Pacific Command at Pearl Harbor, HI as a political-military Intelligence Analyst.

L. Chong (✉)

D.A. Christofferson (ed.), *Women in Security*, Women in Engineering and Science, https://doi.org/10.1007/978-3-319-57795-1_13

Since then, I have earned a M.A. in International Affairs and served at other Department of Defense (DoD) organizations, the U.S. Department of State, and across the Bush and Obama Presidential Administrations at the White House.

Shifting from Cybersecurity and Terrorism to Process Improvement and Project Management

Over the past few years I began to transition away from being a subject matter expert (SME in DoD parlance) in regional security issues, cybersecurity, and counterterrorism, towards business process improvement and project management; I find it exhilarating to enable national security professionals to do their jobs while not working directly on the content.

Coming from a liberal and Liberal Arts background, as well as being multilingual, I find that my way of communicating and creative, holistic, thinking is often different than those around me. My unique background has enabled me to resist groupthink and attack difficult problems such as business process improvement, which pays dividends and adds value to the national security field.

The work we do is essential—threats against the United States are always evolving, amplified by technology and social media. This benefits those who are flexible, willing to take calculated risks with confidence, and proactively embrace change rather than fight it. This flexibility has served me well over the years: because my professional experience is so diversified, I can pivot in a number of directions as project scope or current events necessitate.

Restrictions on Information Sharing

Let me share a vignette from my first job—I attended a 2004 interagency conference on Asia regional security issues, and the final agenda item was on how to more effectively facilitate information sharing between intelligence agencies. There were about 40 SMEs sitting around the room, each representing a military branch of service or intelligence agency; I was the representative from U.S. Pacific Command. U.S. government agencies often work in silos and analysts sometimes do not share information outside their agencies due to restrictions on information sharing, lack of communications infrastructure, or management culture. As you can imagine, this tendency to stovepipe time-sensitive intelligence has significant implications for national security.

The conference chair, a very senior intelligence officer, asked for ideas on how to facilitate interagency information sharing. After no one else volunteered a comment or suggestion, I raised my hand and suggested that we create a blog to share intelligence reporting and coordinate finished intelligence products. The conference chair got red in the face and said the idea "stupid" and it would never happen on his watch. I was stunned at his emotional reaction to a valid suggestion; a tiny voice in my head whispered that I was an imposter and the idea *was* stupid, but I was determined not to take it personally.

After the conference concluded, several representatives from other agencies thanked me for speaking up and said we should absolutely have a blog site to share intelligence information, because there was no way to share information other than by point-to-point emails which is not reliable or sustainable. I kept in contact with these colleagues and we worked with other SMEs and stakeholders to roll out a classified blog with interagency points of contact, a shared documents library, and information on upcoming conferences or events.

While that workplace example is charmingly outdated, this demonstrates the importance of following your instincts and working collaboratively with like-minded colleagues when met with pushback or resistance to innovation.

Groupthink in National Security

Groupthink and lack of diverse ideas is a very real problem in the national security field. I believe one reason is there are not enough millennial innovators who enter the field, commit to a career, and become a senior level decision-maker to enact positive change and innovation. This is frustrating but is balanced by the satisfaction of knowing that by coming from a diverse worldview, utilizing excellent communication skills, you contributed to the mission, and even in a small way, your good decisions help keep men and women and our country safe.

As a minority in this field, I have been closely watched because I stand out, but I always turned that scrutiny into an opportunity to excel. The statistics are well known—women, especially younger women of color—are absent at the top of organizations.

One Door Closes and Another Opens

My sense is that our upbringing does not always train us to succeed in this testosterone-heavy defense industrial complex. For example, the messaging I received growing up from my parents and at school was to behave, get good grades, and don't rock the boat; while compliance was an excellent way to make parents and teachers happy it was not necessarily a recipe for success in the real world.

Being raised to be a "good girl" meant that I have to override my instincts on a near-daily basis and to speak up and be heard. I was slow to become confident in my abilities, have a voice in meetings, and to cease feeling like an imposter in the room.

Over time, successes, failures, and hard work have given me the courage and conviction needed to be successful in this industry and to believe I can contribute in a meaningful way.

Another key factor for success is to ignore the haters—I've had managers tell me I "wasn't a good fit" (every other employee was an older Caucasian male) or that "I didn't seem committed to the organization" (after being promoted without a raise).

Doors will slam in your face but don't be afraid to bang on another one; if an organization does not align with your personal values, don't burn your bridges but do consider finding a new job. I've been told this lack of loyalty is a uniquely millennial mindset—but I believe you should always look out for yourself.

Utilizing Soft Skills to Get Things Done

My niche is enabling others to do their jobs on a project or program—working with stakeholders across unwieldy national security organizations to effect a small or large organizational change is what I enjoy doing.

"Soft skills" such as business process reengineering, strategic communications, and stakeholder engagement are often overlooked; too often large-scale organizational change is implemented without considering the people side of things.

Since the 2004 conference example above, I've found that utilizing "soft skills" effectively had a positive impact on a number of projects such as transitioning an intelligence organization from shared drives and SharePoint to the cloud, or streamlining the delivery of time-sensitive cybersecurity products.

Be Open When Opportunity Calls

Again, there is ample space in the national security industry especially for non-technical staff who can bring a similarly diverse and unique perspective to the field. It is imperative that we attract and broaden the base of women and minorities entering this industry, and also to retain them.

Part IV
Alternative Careers

Chapter 14
Improving Women's Participation in the Security Field

Diane Barrett

Getting Women Involved Through Education and Publication Opportunities

This chapter shares the experiences of the author while exploring why there is a lack of women in security. The theme of the chapter is getting women involved in the field through education and publication opportunities. It includes outreach programs and resources that are available to help those interested in pursuing interests in this ever expanding area of technology.

Few Women in the Field

Looking back to the early days of my IT career, there were very few women in the field. There was a standing joke that the best thing about being a woman in technology was never having to wait in line for the restroom. At lot has changed since then. Women hold high security positions in some of the most well-known technology organizations such as Google and Microsoft.

D. Barrett, Ph.D., C.I.S.S.P. (✉)

D.A. Christofferson (ed.), *Women in Security*, Women in Engineering and Science, https://doi.org/10.1007/978-3-319-57795-1_14

Career Beginnings

I enjoyed math and originally went to college for a degree in accounting. I became an IRS enrolled agent but I was bored sitting in an office for eight hours every day. I discovered that I really liked working with computers. I began my technology career working for a software development company. Learning about Windows NT and integrating our proprietary DOS based software for proper performance on NT was quite the accomplishment. Additionally, installing the software on customer systems required traveling to the customer site, so there was ample opportunity to see what types of networks existed in other organizations.

Education and Certifications

During this time, education and certifications became an important part of a career in technology. It was sometimes difficult to work with other administrators and vendors without having credentials. Certifications came first because they were easier to obtain than a degree and as experience increased, so did the likelihood of passing certification tests.

Technical College Instructor

Tired of traveling and being constantly on call, I accepted a position as an instructor at a local technical college in the Computer Networking Technologies program. In conjunction with studying for additional certification exams, there was opportunity to do technical editing for a local company that published ExamCram books. The contacts at ExamCram also afforded additional visibility as an author and instructor for short technology courses for various companies including HP and Forbes.

Writing Security Books

As networks began being interconnected, security became a topic of concern. I first became involved in security and then digital forensics. While searching for material to study for the CompTIA Security + beta exam, a posting seeking authors for the Security + ExamCram was answered. This began a writing career that continues to this day. The Security + ExamCram SYO-401 edition, which is the fourth iteration of CompTIA's Security + exam, went to publication in 2015. The same three original authors have done all four books. This is a great feat in itself as it often very difficult to work with other people on a writing project. I have written over a dozen books related to technology, security and forensics.

Academia, Industry and Educational Rigor

I enjoyed my time in education, but missed the field work. For quite a number of years I did both. At first, I worked in industry part-time and in education full-time. Then I worked in education part-time (mainly online) and in industry full-time. During this time, I advanced my education up to the point of completing a PhD in business administration with a specialization in information security. This type of educational rigor is not for the light of heart, but due to all the prior writing work I did, the dissertation process went a bit smoother than for those that did not have the pleasure of previously having someone rip apart their work.

Shared Roles in Industry and Education

For now, my work consists of working in industry part-time and spending most of my time in education. At this point in my life, it is a bit more enjoyable to have some flexibility and not have to travel on a moment's notice. I have presented at a variety of conference venues, served on boards and panels, chaired conferences, and am associated with many industry organizations.

Why the Number of Women Is So Low in Security

With the background information complete, it's time to move on to explore the crux of this chapter, discussing why there is a significantly low number of women in security and improving female involvement in the field through education and publication opportunities. A Cisco report estimates that there are one million global cybersecurity job openings (Cisco 2015). By 2019, the demand is expected to reach six million globally, with a projected shortfall of 1.5 million (Morgan 2016a, b). Traditionally, technology fields have been primarily male dominated. The Women's Society of Cyberjutsu (WSC) states that only 11% of the world's information security workforce are women (Cooke 2016).

The lack of women in the security workforce is attributed to several factors that include unfriendly work environments, family responsibilities, and complex job demands (Georgetown University 2016). Hill et al. (2010), presented eight research findings from the results of a NSF grant for the American Association of University Women (AAUW). The study drew on a large and diverse body of research, providing evidence that social and environmental factors contribute to the underrepresentation of women in science and engineering. One of the report recommendations was that encouraging more girls and women to enter vital fields such as security, will require careful attention to the environment in our classrooms, workplaces and culture. Since I have a direct impact in these areas, I hope to make a difference by encouraging more women to enter the field of security through sharing my experiences and helping them be successful.

The low percentage of women in security does not truly reflect the current state of the security profession. The number of women employed in the certain areas of security has actually increased at a greater pace than that of men. For example, in the most recent Frost & Sullivan ISC2 survey, one in five women is in a governance, risk, and compliance (GRC) role compared one in eight for men (Frost & Sulluvan 2015). The difference was attributed to women pursuing and seizing new opportunities that were emerging in GRC to a greater extent than men. The study also revealed that the percent of women with either a Master's or Doctorate degree exceeds the percent of men in both practitioner and leadership roles. In leadership roles, 58% of women had advanced degrees versus 47% of men (Frost & Sulluvan 2015). Based on the figures provided, it can be deduced that the security profession not only attracts, but requires individuals of high academic achievement. The report recommendation was that organizations promote the profession by supporting cybersecurity education in primary schools, offering internships, pairing new hires with mentors, and adapting compensation plans and training to better align with flexible working arrangements.

A Network of Contacts and Connections

Most seasoned industry professionals have built a network of contacts and connections. For those just entering the industry, finding a job or making contacts can seem a bit intimidating. There are ample opportunities for networking and industry contacts depending on the direction one wishes to take. In addition to in-field job leads there are additional careers that can be explored such as publishing and teaching. This quote from Yogi Bhajan touches on the value of writing and teaching: "*If you want to learn something, read about it. If you want to understand something, write about it. If you want to master something, teach it.*"

Sharing Industry Knowledge Through Teaching

Various venues are available for sharing industry knowledge through teaching. Many professionals enjoy teaching part-time while others take the plunge into a full-time position. Numerous colleges have open positions posted for faculty in information security programs. Community and technical colleges are more likely to accept part-time faculty without an extensive educational background. Public institutions are often bound by accreditation requirements and may allow an industry professional to teach at one level lower than the highest degree held. For example, someone with a master's degree can teach at the bachelor level. Full-time educational employment usually means that the person is either working toward or already has an advanced degree such as a PhD. Private colleges are different in this respect. Many look for industry experience as opposed to educational background. One of the best elements of working in education is that more and more institutions are offering online programs. Teaching in online programs affords one a lot of flexibility as well as the opportunity to still work in the field. I spent several years teaching only online and really enjoyed being able to adjust my schedule at a moment's notice without affecting the classroom experience.

Publication Adds Credibility and Visibility

Publication is a way to add credibility and visibility to one's profile. Publication opportunities are varied and include books, journals, and white papers. Book publishers such as Jones & Bartlett Leaning are always open to suggestions for technical books. This publisher in particular has quite a few security related books. Technical writing for publication is a bit different than writing a best seller. Contrary to popular belief, there is not a lot of money in technical book writing. For example, one Amazon reviewer believed that my book royalties were being used to pay for classes while I worked on my PhD. If only writing a technical book was that lucrative. Most people publish for the visibility, not the money. Advances are small and royalties usually aren't paid for several years.

That's not to say there isn't money to be made if you enjoy the solitude of spending your day researching and typing. In fact, a literary agent will procure writing contracts for you, usually for a percentage of the royalties. I have published work that is under a literary agent contract and some that is not. Most of the agent related work originated early on from an established agency partnership. Literary agencies such as Waterside Productions, Inc. serve non-fiction authors and publishers for print and digital media. Agents generally help keep the project on track, negotiate contract terms, and issue royalty payments.

Writing a book is an experience that may leave one ambivalent. It is great to see your name on the cover, but sometimes by the time you get there, you never want to repeat the experience. Book projects tend to be multi-author, allowing the book to get to market in a timely fashion. If all authors meet the deadlines, the project tends to go well. If an author does not meet deadlines or is unresponsive, the project doesn't go so well. When an author becomes unresponsive, the project falls behind and sometimes the contract has to be renegotiated. All book publisher websites have information on how to publish a book and some offer help as well as insight into writing a book. Elsevier SciTech Connect has a blog page titled "*What is it like to publish your first book?*".

In some instances, there is no payment at all to authors especially when it is part of a large project. Academic publishers such as IGI Global require the submission of a proposal form outlining the details of the book idea. The book proposal is reviewed and accepted, the submitting editor does a call for chapters, and then numerous authors respond to the call for chapters. Once your chapter is written, it is submitted, peer reviewed, revised, and then resubmitted. As an author, this type of project is good because once your final chapter is submitted, you are done with the project until the final publication is released. These publications are almost always marketed internationally, giving the author's work exposure to a wide audience. I just completed a chapter for The Encyclopedia of Information Science and Technology, 4th edition. This work is a 10-volume major reference work for which over 900 papers were submitted for publication.

Journal publication is less intensive than writing a book or book chapter. This type of publication tends to be a bit more academic. Most journal submissions are double-blind peer reviewed and the submission structure tends to be based on some type of research. Academic employment requires publication of this type simply because the work is reviewed by peers.

The last type of publication is a white paper. A white paper is a technical or business benefits document that introduces a current challenge, then makes a strong case why a particular approach to solving the challenge is preferred (Stelzner 2007). Generally, white papers describe a solution to a problem, but may also outline how to perform technical tasks or introduce a new concept.

White papers can be published on vendor websites, professional organization websites, or more formal location such as the Cornell University library. Publishing a white paper on a professional organization website may have other advantages. For example, if a Digital Forensic Certified Practitioner (DFCP) publishes a white paper and posts it to the DFCB website, credit is given toward the member's recertification requirements.

Many Ways to Be Involved in Security Without a 9-5 Office Job

As one can see, there are many ways to be involved in the information security field and not all require working a regular 9 to 5 job. I do better in a position where it is not mandatory to spend 40 h in an office environment and there is a reasonable amount of flexibility. The career choices I made have allowed me to be successful in some of the more non-traditional, yet vital aspects of information security.

This last section contains a few final thoughts, followed by resources that are available to help those interested in pursuing interests in this ever expanding area of technology.

How to Increase the Number of Women in the Field

When we involve girls at an early age, they are more likely to pursue a career in the field. A key factor in predicting STEM career interest at the end of high school was interest at the start of high school. Early exposure to information security that sparks an interest in the field is often a precursor to a girl actually pursuing a career in the field (Corbett and Hill 2015).

Encouraging girls to enter and stay in the field, will help close the gender and employment gap in information security positions and help strengthen our country's cybersecurity posture.

Women stress the need to look beyond technical skills in the hiring process for information security jobs. Technical skills are an important part of the job, but technical skills alone are inadequate for resolving the complex risk management problems that leaders in information security face on a regular basis (Frost & Sulluvan 2015).

The world of information security is vast and constantly changing. We have the power to make a difference not only for ourselves, but also for the women that will follow in our footsteps.

Resources for Women in the Field

Crystal Bedell, a freelance technology writer, posted these top resources for women in technology to the IT Job Cafe blog in September of 2015:

- Anita Borg Institute (ABI)—ABI seeks to help women in computing reach their career goals by providing opportunities to learn, network with other women and stay inspired
- The National Center for Women & Information Technology—NCWIT is a non-profit community that seeks to increase the number of women working in computing and technology
- Women Who Code—WWCode is a US-based non-profit dedicated to inspiring women to excel in technology careers
- Women in Technology International—WITI is a trade association for women who use technology in any job function, including finance, human resources, marketing, management, sales, and IT.
- TechWomen—TechWomen is an initiative of the U.S. Department of State's Bureau of Educational and Cultural Affairs to help strengthen relations between the U.S. and the Middle East and North Africa.
- STEMinist—STEMinist was created in 2010 by Ann Hoang to increase the visibility of women in the fields of science, technology, engineering and mathematics (STEM)

These resources were recommended on Forbes Tech Blog in March of 2016:

- SANS CyberTalent Immersion Academy for Women—This accelerated training and certification program offers women a fast track to top jobs in cybersecurity
- The Women's Society of Cyberjutsu—(WSC) is a 501(c)3 non-profit passionate about helping and empowering women to succeed in the Cybersecurity field
- The Women in Cybersecurity Project—As part of New America's Cybersecurity Initiative, the project brings together cybersecurity companies, government, and big thinkers to promote methods to bring women into the cybersecurity field
- Women in Cybersecurity—WiCyS—brings together women in cybersecurity from academia, research, and industry for sharing of knowledge, experience, networking and mentoring

References

Bedell C (2015, September) Top resources for women in technology. Retrieved from http://www.itjobcafe.com/blogs/Top-Resources-For-Women-in-Technology/130

Cisco (2015) Mitigating the cybersecurity skills shortage. Retrieved from http://www.cisco.com/c/dam/en/us/products/collateral/security/cybersecurity-talent.pdf

Cooke L (2016, April) In focus: the desperate shortage of women in cyber security, endpoint security news. Retrieved from http://solutionsreview.com/endpoint-security/focus-the-desperate-shortage-of-women-in-cyber-security/

Corbett C, Hill C (2015) Solving the equation: the variables for women's success in engineering and computing. AAUW, Washington, DC. Retrieved from http://www.aauw.org/aauw_check/pdf_download/show_pdf.php?file=solving-the-equation

Frost & Sulluvan (2015) The 2015 (ISC)2 global information security workforce study. Retrieved from https://www.isc2cares.org/uploadedFiles/wwwisc2caresorg/Content/GISWS/FrostSullivan-(ISC)%C2%B2-Global-Information-Security-Workforce-Study-2015.pdf

Georgetown University School of Continuing Studies (2016) Women in tech thrive in a male-dominated field. Retrieved from http://scs.georgetown.edu/departments/15/master-of-professional-studies-in-technology-management/news/5206/women-tech-thrive-male-dominated-field

Hill C, Corbett C, & St Rose A (2010) Why so few? Women in science, technology, engineering, and mathematics. American Association of University Women, Washington, DC. Retrieved from https://www.aauw.org/files/2013/02/Why-So-Few-Women-in-Science-Technology-Engineering-and-Mathematics.pdf

Morgan S (2016a, March) Calling all women: the cybersecurity field needs you and there's a million jobs waiting. Forbes Tech. Retrieved from http://www.forbes.com/sites/stevemorgan/2016/03/28/calling-all-women-the-cybersecurity-field-needs-you/#698dfffc5ca4

Morgan S (2016b, January) One million cybersecurity job openings in 2016. Forbes Tech. Retrievedfromhttp://www.forbes.com/sites/stevemorgan/2016/01/02/one-million-cybersecurity-job-openings-in-2016/#214b0147d274

Stelzner MA (2007) Writing white papers: how to capture readers and keep them engaged. WhitePaperSource, Poway

MIX
Papier aus verantwortungsvollen Quellen
Paper from responsible sources
FSC® C105338

If you have any concerns about our products,
you can contact us on
ProductSafety@springernature.com

In case Publisher is established outside the EU,
the EU authorized representative is:
Springer Nature Customer Service Center GmbH
Europaplatz 3, 69115 Heidelberg, Germany

Printed by Libri Plureos GmbH
in Hamburg, Germany